PRAISE FOR DR. JOE AND WHAT YOU DIDN'T KNOW:

"Mother Nature has a few less secrets. In *Dr. Joe and What You Didn't Know*, Dr. Joe Schwarcz enlightens us with explanations for life's curiosities."

— Joy Ferguson, *Canadian Bookseller*

PRAISE FOR THAT'S THE WAY THE COOKIE CRUMBLES:

"Schwarcz explains science in such a calm, compelling manner, you can't help but heed his words. How else to explain why I'm now stir-frying cabbage for dinner and seeing its cruciferous cousins—broccoli, cauliflower, and brussel sprouts—in a delicious new light?"

— Cynthia David, *Toronto Star*, online

PRAISE FOR THE GENIE IN THE BOTTLE:

"Often lighthearted, other times deadly serious, he covers a range of topics, 64 in all, that includes tap water quality, methane, bag balm, and even 'flubber.' No doubt readers will find a lot of stuff here about a lot of things they have never even considered."

— Ted Hainworth, *Star Phoenix*, 29 December 2001

PRAISE FOR RADAR, HULA HOOPS, AND PLAYFUL PIGS:

"Dr. Schwarcz has written a book that has done three things which are difficult to do. First, the book is enormously enjoyable — it commands and holds your attention. Second, it explains science and scientific phenomena in a simple and yet accurate way. And third, it stimulates you to think logically and in so doing, it will lead to a scientifically literate reader who will not be so easily misled by those who wish to paint science and technology as being a danger to humankind and the world around us."

— Michael Smith, Nobel Laureate

"It is hard to believe that anyone could be drawn to such a dull and smelly subject as chemistry — until, that is, one picks up Joe Schwarcz's book and is reminded that with every breath and feeling one is experiencing chemistry. Falling in love, we all know, is a matter of the right chemistry. Schwarcz gets his chemistry right, and hooks his readers."

— John C. Polanyi, Nobel Laureate

THE FLY IN THE OINTMENT

THE FLY IN THE OINTMENT

70 Fascinating Commentaries
on the Science of Everyday Life

DR. JOE SCHWARCZ

Director
McGill University Office for Science and Society

ECW PRESS

Published by ECW PRESS
2120 Queen Street East, Suite 200, Toronto, Ontario, Canada M4E 1E2

NATIONAL LIBRARY OF CANADA CATALOGUING IN PUBLICATION DATA

Schwarcz, Joseph A.
The fly in the ointment: 70 fascinating commentaries on the science of everyday life / Joe Schwarcz

ISBN 978-1-55022-621-8

1. Science – Popular works. I. Title.

Q173.S38   2004   500   C2003-907310-6

Copyeditor: Jodi Lewchuk
Production: Emma McKay
Interior design: Yolande Martel
Interior cartoons: Brian Gable
Cover design: Guylaine Regimbald – SOLO DESIGN
Cover artwork: © 2003 Varian (www.varian.net)
Author photo: Tony Laurinaitis
Printing: Transcontinental

Second Printing

This book is set in Stempel Garamond and Koch Antiqua.

The publication of *The Fly in the Ointment* has been generously supported by the Canada Council, the Ontario Arts Council, and the Government of Canada through the Book Publishing Industry Development Program.

Canada   Canada Council / Conseil des Arts   ONTARIO ARTS COUNCIL
          for the Arts   du Canada   CONSEIL DES ARTS DE L'ONTARIO

DISTRIBUTION
CANADA: Jaguar Book Group, 100 Armstrong Avenue, Georgetown, Ontario L7G 5S4

UNITED STATES: Independent Publishers Group, 814 North Franklin Street, Chicago, Illinois 60610

EUROPE: Turnaround Publisher Services, Unit 3, Olympia Trading Estate, Coburg Road, Wood Green, London N22 6TZ

AUSTRALIA AND NEW ZEALAND: UNIREPS University of New South Wales, Sydney, NSW, Australia 2052

PRINTED AND BOUND IN CANADA

ECW PRESS
ecwpress.com

CONTENTS

INTRODUCTION

THE FLY IN THE OINTMENT

Many of you may not remember the ditto machine, but I sure do. I don't know what I would have done without it back in 1973 when I first started teaching. It offered the only quick, cheap way to produce handouts for students. Reproduction required a stencil, which was made of a master sheet attached to a backing sheet treated with purple ink. The pressure created by writing or typing caused the ink to be transferred to the back of the master, which would then be peeled off and wrapped around the metal drum of the duplicating machine. This drum was connected to a container of methanol by a wick, and as the drum turned the methanol softened the ink and transferred it onto blank paper. Several dozen copies could be produced before the ink ran out.

I mention this anecdote because I vividly recall the first such stencil I ever produced. I had just graduated with a Ph.D. in chemistry and was keen on communicating my knowledge to students. The only problem was that I didn't have very much knowledge to communicate. Or at least not the type I thought I should be communicating. Oh, I was pretty good at atomic structures, chemical bonding, reaction mechanisms, and even thermodynamics—all fundamental concepts that any chemistry student should master. But I knew precious little about "real-world" chemistry. That had not been part of my chemical education. I knew how to interpret complex spectra but had no idea why carrageenan was added to ice cream. I knew how to make carbon 13–enriched glucose in the lab but would have been totally stymied if someone had asked me to make lipstick. I discovered that when friends and relatives learned I was studying chemistry, they were more likely to ask me questions about toothpaste or shoe polish than about the nuances of the Schrödinger wave equation.

I decided that when I finally got the chance to teach, I would always try to weave these everyday applications into my courses. Luckily, my very first year on faculty I had the opportunity to develop a new course that was to feature dyes, cosmetics, cleaning agents, medications, synthetic fabrics, and the other common fruits of chemistry. These were just the kinds of things I was interested in, but curiously, "Chemistry in the Modern World" was to be offered only to nonscience students! The pedagogical mentality at the time suggested that such "fluff" was fine for arts students, but there was no room for it in "real" chemistry courses. Science students were to struggle with phase diagrams and molecular orbitals, not with ways to remove lipstick stains from collars. My view was that these real-world examples should be part and parcel of any chemistry course.

Nevertheless, I was thrilled to be able to teach my arts students about stuff I thought really mattered. I thought they should learn about suntan lotions, preservatives, chemotherapy, shampoos, and air pollution. I thought they should learn that chemistry is a living science and that there is always some "breaking news." So I took to starting each class with a "ditto" handout about some current chemical happening. One day it might have been about the benefits of a newly introduced medication, and the next it might have mentioned the wonders of a novel plastic. The more I pursued this practice, the more I became captivated by the scope of chemistry. But something else happened as well: I began to realize that there is always a "but." That new drug may perform well most of the time, but sometimes there are severe side effects. That new plastic may have fantastic properties, but there are environmental concerns linked to its production.

In other words, I discovered that there is often a "fly in the ointment." It became clear to me that any realistic discussion of chemical issues had to involve an appropriate risk-benefit analysis. And that is just the approach that my colleagues, Ariel Fenster and David Harpp, and I now take as we offer applied chemistry courses to over 1,000 students from all disciplines at McGill University in Montreal, Canada, every year. In fact, we have given these four courses to almost 14,000 students over the past twenty-four years. Much has changed since my initial attempts to offer a relevant chemistry course in 1973. Today, our lectures are available on the Web <http://www.oss.mcgill.ca> and "ditto" handouts have been relegated to the dustbin of technology. But I still recall those old dittos with a certain degree of fondness—after all, they did spark many a fruitful discussion. I also remember, though, the headaches I used to get from the methanol vapors when I was running off all those handouts. You see, there was a fly in that ointment as well. So as you journey through the following pages, you too may encounter some "flies." You'll have to judge for yourself just how much they contaminate the ointment.

HEALTH ISSUES

LIES, DAMNED LIES, AND STATISTICS

I overheard an interesting conversation between two young women as I was waiting in line to ride The Comet, the grand-daddy of roller coasters at Great Escape Fun Park in upper New York State. One was preparing to study in Australia and was describing her travel plans. Her friend thought Australia would be exciting but added that she would never go herself because flying was too dangerous. The prospective traveler responded that she wasn't concerned about a plane crash but was worried about the risk of developing deep vein thrombosis—a potentially fatal blood clot—during the trip. As soon as I heard this comment I knew that she must have watched the same TV talk show I had the day before.

While the specifics of this conversation might have been unique, the gist of it was not. Details aside, the young women were involved in risk evaluation, something we all do on a regular basis. Just think about how often we ask ourselves whether or not we should be worried about mercury in tuna, radiation from cell phones and microwave ovens, aspartame in diet drinks, and the reported link between estrogen supplements and an increased risk of cancer. Life often comes down to analyzing risks. But most people do not realize how difficult it is to perform this analysis in a meaningful way. Let's start with something easy, like the risks of air travel. Flying is actually remarkably safe. Since the advent of commercial air transport around 1914, some 15,000 people have perished in airplane crashes. In North America alone more than three times that many people die in automobile accidents every year! You are far more likely to arrive at your destination if you fly than if you drive. Unfortunately, traffic on the highways increased significantly after September 11, 2001, resulting in many deaths that would not have occurred if people had flown. So why are people so scared of flying? Because they don't think statistically—they think emotionally. People have the perception that their destiny is in their own hands if they are driving a car; they feel that they have relinquished this control when flying on an airplane. Also, there is a greater likelihood of surviving a car crash than a plane crash, which is another factor that weighs on people's minds. But the statistics show that over a lifetime, you are 100 times more likely to die in a car accident than in a plane crash. Basically, you are more likely to be struck by lightning or win the lottery than die in an air disaster. Flying to Australia is safer than driving to the Great Escape from New York City, which is what the two young women had done.

Now, about the deep vein thrombosis. The TV show I had seen focused on what has come to be called "economy-class syndrome" and began by recounting the tragic case of a healthy British woman in her late twenties who collapsed at Heathrow Airport in London after a long flight from Australia. She died within hours from a blood clot in her lung that had originally formed in her leg while she sat in a cramped position for an extended period. The show also included interviews with physicians in Hawaii who described similar incidents. There is no question that such deep vein thrombosis can occur, but the number of people who develop this condition is very small when compared with the number of passengers that fly. Indeed, a study reported in the *New England Journal of Medicine* found no association between air travel and deep vein thrombosis. On long flights, passengers should certainly be encouraged to move around, particularly if they are seniors, have a history of heart disease, are pregnant, or are taking estrogen supplements. The risk of deep vein thrombosis on a flight is a minute statistical blip, but if you take three victims and put them on a talk show together, viewers will think that the air travel industry is in midst of an epidemic.

If you want something to worry about on a flight, worry about the air quality. Although the recycled cabin air is filtered and mixed with fresh air, during flight many passengers complain of nausea and flu-like symptoms, which are consistent with a reduced oxygen supply. Of even greater concern is the spread of infectious organisms within the confines of an airplane cabin. In one documented case, passengers were kept on board while a plane underwent a minor repair. One of the passengers had a case of influenza A, which spread to three-quarters of the other travelers.

The way data are communicated can also affect people's perception of risk and the decisions they make. Take, for example, the recent study that showed a 30 percent increase in risk of breast cancer among women taking estrogen supplements. Sounds terribly frightening! But consider that within a ten-year period, about 3 to 4 percent of menopausal women will be struck by breast cancer. A 30 percent increase in this risk means that that number rises to 5 percent; suddenly the 30 percent increase in risk doesn't seem quite as impactive. Putting it another way, a postmenopausal woman who takes estrogen reduces her chance of remaining cancer-free from about 96 percent to 95 percent.

Consider also this example: A pharmaceutical company's ad, aimed at physicians, touted a drug's ability to reduce the risk of heart attack in a group of patients by 25 percent. In the wake of the ad, the number of prescriptions being written shot up. In reality, however, the drug study showed that 0.4 percent of patients had fatal heart attacks when not taking the drug as opposed to 0.3 percent of those who did take it. When statisticians examined the data they concluded that if more than seventy patients were treated with the drug for five years, the treatment would help just one of them avoid a heart attack. With results stated in this way, the drug sounds far less appealing.

And how about this: Benzopyrene in charcoal-broiled foods is a known carcinogen, because in large doses it will cause cancer in rodents. Based on the animal model, scientists estimate that eating 100 charcoal-broiled steaks will increase the risk of cancer by 1 in 1 million, a number that has arbitrarily been selected as a "red flag." What does this statistic really mean? Since the risk of cancer over a lifetime is about 1 in 3, eating these steaks will raise that risk by 0.0003 percent. It makes more sense to limit consumption of steaks because of their fat content, not because "they cause cancer."

It would have been fun to engage the two young women at the Great Escape in a discussion of risks. I could have told them about the sixty-four-year-old man who developed "shaken baby syndrome" after riding a roller coaster. Given that the two were smoking like steam engines, I also would have liked to mention that every 1.4 cigarettes increased their risk of cancer by 1 in 1 million. But with all that secondhand smoke around, I didn't want to venture any closer. Risk analysis is a risky business.

FARMED, WILD, OR CANNED?

It's a pretty common scenario these days: Scientists publish a paper about the presence of a synthetic pollutant in a consumer product and warn people about excessive exposure because the substance is known to cause cancer or reproductive problems when fed to rodents in high doses. Said study about yet another cancer-causing substance in the environment makes front-page news. Spokespeople for the industry in question complain bitterly that the risk has been exaggerated while environmentalist groups hail the study as a breakthrough. Scientists with impeccable credentials wade into the debate on both sides, sometimes accusing each other of having vested interests. Different government regulatory agencies can't agree on what recommendations to make. The public is thoroughly confused, and my office gets lots of e-mails and phone calls.

A prime example of this scenario is a scare triggered by a paper published in the prestigious journal *Science* in 2004. In it, researchers reported that farmed salmon are significantly more contaminated with organochlorine compounds such as PCBs, dioxins, toxaphene, and dieldrin than their wild counterparts. PCBs were once commonly used as insulating fluids in electrical equipment, dioxins are by-products of some industrial processes, and toxaphene and dieldrin are insecticides. These chemicals are particularly persistent in the environment and are fat-soluble. As a result, they accumulate in the fatty tissues of farmed fish, which are fed fish meal and oil made from smaller fish that also contain deposits of these chemicals. Similarly, when we eat contaminated fish, the organochlorides can build up in our fatty tissues. Everyone agrees that this buildup is not a good thing. Why? Because there is evidence that these compounds are capable of producing some pretty nasty health effects.

Let's use PCBs as an example and examine the related cancer

risk. There is no question that PCBs can cause the disease in animals, with the liver being the main organ affected. The human picture is less clear. Epidemiological studies have shown that workers with extensive exposure to PCBs in an industrial setting suffer a slightly elevated risk of cancer. Some investigators have also found a significant association between PCB concentrations in fatty tissue and non-Hodgkin's lymphoma. A couple of incidents in Japan and Taiwan, in which people ingested rice oil accidentally contaminated with a high dose of PCBs, are also suggestive of an increased risk of liver cancer. Labeling PCBs as probable human carcinogens therefore seems justified. As we have seen, PCBs are present in fish, particularly in the farmed variety. But that does not mean that eating fish raises the risk of cancer. Our food supply contains numerous carcinogens, both natural and synthetic. Hydrazines in mushrooms, heterocyclic aromatic amines in cooked meat, aflatoxins in molds, and acrylamide in baked goods are all carcinogenic. But our diet also contains anticarcinogens in the form of various vitamins and polyphenols. When we eat we consume hundreds of different chemicals, and the result of their interplay in our body is virtually impossible to predict. That's why the appropriate question to ask is not whether organochlorine contaminants in fish can cause cancer, but whether a diet high in fish can do so. I am unaware of any study that shows a link between increased fish consumption and cancer. On the other hand, numerous studies point to just the opposite conclusion!

Swedish researchers have clearly shown that eating fatty fish, salmon in particular, can reduce the risk of prostate cancer by one-third. Italian and Spanish scientists have investigated the relationship between frequency of fish consumption and cancer and found that for those who consumed fish regularly, there was a consistent pattern of protection against the risk of digestive tract cancers, particularly of the colon, one of the leading causes

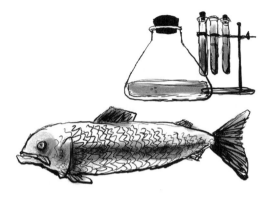

of cancer mortality in developed countries. At the Aichi Cancer Centre Hospital in Japan, scientists looked at the diets of more than 4,000 healthy people and another 1,000 with lung cancer. Both men and women who ate large amounts of fresh fish were significantly less likely to develop lung cancer. This finding may explain why the Japanese, who smoke more than Westerners, have a lower rate of lung cancer. An extensive survey conducted over ten years, involving more than 60,000 people of Chinese descent in Singapore, found that women who eat at least 40 grams of fish per day reduced their risk of breast cancer by 25 percent. There is sound theoretical justification for these observations. Prostaglandins are a class of chemicals in the body that produce a variety of hormone-like effects, some of which are linked to carcinogenesis. These chemicals are derived from arachidonic acid, which in turn is formed from linoleic acid, a common omega-6 fat in the diet. Fish oils inhibit the cyclooxygenase-2 enzyme that converts arachidonic acid to the problematic prostaglandin E2. Essentially, reducing fish intake is likely to result in more—not less—cancer, irrespective of the contaminants fish may contain.

While the prospect of cancer instantly strikes fear into the heart, the fact is that strokes and heart disease kill more people. And there is overwhelming evidence that links fish consumption with protection from strokes and heart attacks. But why stop with cancer, strokes, and heart disease? Recent evidence indicates that fish consumption offers protection from diabetes and maybe even from Alzheimer's disease. In all cases, the beneficial chemicals are believed to be the omega-3 fats, of which salmon may be the richest source. In fact, farmed salmon, on average, contain somewhat more omega-3s than their wild counterparts. Furthermore, salmon—farmed or wild—is less likely to be contaminated with mercury than other commonly eaten fish. So what you want to ask yourself is whether you should put more emphasis on the theoretical risks associated with organochlorides in fish or the proven benefits of fish consumption.

Although the answer to this question should be obvious, the salmon study in *Science* is still an important one. It will undoubtedly encourage fish producers to take steps to reduce the organochloride residues in their product, something that is technically feasible. The use of feed made from canola and soy oil genetically modified to contain more omega-3 fats is an interesting possibility. Incidentally, canned salmon almost always comes from wild Alaskan salmon, which are minimally contaminated with organochlorides. Most fish oil supplements, of which the usual recommended dose is 2 to 4 grams per day, are also free of these compounds.

While the authors of the *Science* paper deserve credit for compiling valuable data on salmon contamination, I believe their argument—which is that more than one meal of farmed salmon a month may hike the risk of cancer—is totally unjustified. Indeed, I think it can be effectively argued that any such cutback in salmon consumption is seriously detrimental to health. Since wild salmon is quite expensive, the warning about farmed salmon could have the effect of significantly reducing salmon consumption in the population, thereby increasing the risk of illness. I would agree, though, that it is a good idea for pregnant women to stick to wild salmon, just to be ultrasafe.

As far as I'm concerned, based on the studies I examined while looking into this issue, I would happily keep eating a couple of servings of fish per week—farmed or wild. Alas, I'm allergic to fish, so I will have to stick to my flaxseed, which is a far poorer source of omega-3 fats than salmon. I'm also told that it doesn't taste nearly as good.

DDT: A Double-Edged Sword

I remember being quite taken when, as a student, I read Rachel Carson's 1962 epic *Silent Spring*. As a biologist, Carson made a compelling case against the synthetic pesticides that had been introduced in the post–World War II era. She maintained that they were responsible for fish kills, pollution of the soil, and reproductive problems in birds. DDT in particular caused thinning of egg shells and led to fewer hatchings. Ospreys, peregrine falcons, and eagles were disappearing, Carson said, and robins were being killed in misguided attempts to eradicate Dutch elm disease by spraying trees with DDT. That's why there would eventually be no birds to sing: there would be a "silent spring."

I was impressed by Carson's book. I thought it was an excellent example of how we cannot always predict the consequences of a chemical intervention and how the introduction of a substance into the environment, although seemingly for all the right reasons, can backfire. Carson made an impassioned plea against putting blind faith in technology, particularly when it came to pesticides such as DDT.

This notorious compound was first synthesized in 1874 by Othmar Zeidler, who combined chloral (which later became known as a "Mickey Finn" after the Chicago bartender who supposedly used it to put his rowdy patrons to sleep), chlorobenzene, and concentrated sulfuric acid to make it. Zeidler was simply interested in making novel compounds for his Ph.D. thesis and never studied DDT further. But in 1939, Paul Muller, working for the JR Geigy Company in Switzerland, did. He was interested in moth repellants and had come across a compound called "diphenyltrichloroethane," which was somewhat effective. Muller then did a literature search and came upon DDT, a closely related substance. He synthesized it according to Zeidler's recipe and discovered that it was remarkably toxic to insects. And much to his satisfaction, it seemed not to have any effect on domestic animals or humans. Swiss farmers were thankful. Just a year after Muller's discovery, DDT was used to wipe out the Colorado potato beetle, which had threatened the country's potato crop.

By 1945 DDT was being used worldwide on numerous crops. But concerns arose with two discoveries: the chemical's application caused it to disperse into the air and spread far and wide, and it was showing signs of toxicity in frogs and fish. By the 1950s it was apparent that DDT was building up in the fatty tissues of animals and humans. Eventually, the US Environmental Protection Agency (EPA) stepped in and banned the substance. Rachel Carson had played her role, the environmental movement had begun, and a major problem had been eliminated.

I vividly recall telling this story in class when I first started teaching back in 1973. I thought Rachel Carson had done a great job. True, I had seen references to the use of DDT during the war to wipe out mosquitoes that transmitted malaria, but frankly, I didn't pay much attention to that. After all, we didn't have malaria in North America. It never occurred to me that maybe it was because of DDT use. After all, Rachel Carson had made DDT out to be a chemical villain, and the EPA had agreed with her. Being young and somewhat naïve, I didn't think to check out some of Rachel Carson's "facts." When I finally did look into the DDT issue more deeply, I began to realize that the picture Carson had painted was not completely accurate. DDT had another side.

When the Germans retreated from the Italian city of Naples during World War II, they dynamited the city's water system. The inhabitants had no water to wash with, and body lice proliferated. The result was an outbreak of *Typhus bellicus,* or "war typhus," a disease that in previous wars had killed millions. This time, though, the Allies had an answer. They had DDT. About 1.3 million Neapolitans were dusted with mixture of talcum powder and DDT, and within three weeks the epidemic was stopped in its tracks. But that was only the beginning. DDT turned out to be highly effective against mosquitoes that transmitted malaria. Sprayed on the walls of houses in the tropics, it would keep the insects away for weeks. In Ceylon (now Sri Lanka), where about 2.5 million cases of malaria were recorded annually in the 1950s, regular spraying led to just 31 reported cases in 1962. The world had never seen such a miraculous result.

Coincidentally, 1962 also marked the publication of Rachel Carson's *Silent Spring,* in which she described DDT as the "elixir of death." And she wasn't referring to insects. Carson was convinced that DDT and other similar pesticides had unleashed a catastrophic plague on the world. In addition to her prediction that wildlife would be affected to the point where no birds would be left to sing in the spring, she claimed that the accumulation of persistent DDT in the bodies of mammals would cause cancer rates to soar. She was certainly right about the persistence and accumulation of DDT. Both it and its major metabolite, DDE, persist in the environment for many years. They are essentially insoluble in water but are very soluble in fat, which means that they accumulate in fatty tissue and build up in the food chain. While plankton in water may have very little DDT, the fish that eat the plankton will have more, and birds that eat the fish more yet. We all have some DDT in our flesh that can be traced back to the massive spraying of agricultural fields and the vast amounts used in insect control efforts prior to 1972, the year when most uses of DDT were banned in North America. Indeed, in 1962, 80 million kilograms of DDT were used worldwide. Carson was absolutely correct when she said that DDT could be found in mountain lakes, in the bodies of polar bears, and at various sites far removed from where it was applied. But its presence is not enough to condemn it as a criminal. What other evidence did Rachel Carson have? Not much.

Silent Spring is dedicated to Albert Schweitzer and his quote that "man has lost the capacity to foresee and to forestall and will end by destroying the Earth." Carson's implication was that the famous physician was speaking about DDT. Actually, as he makes clear in his autobiography, Schweitzer was a proponent of DDT to control malaria, and the quote refers not to DDT but to nuclear warfare. As it turns out, this presentation of Schweitzer's quote out of context is not the only instance in which Carson played loose with the facts.

She referenced a paper by James DeWitt in the *Journal of Agricultural and Food Chemistry* to support her thesis that a decline in certain bird populations was due to the thinning of egg shells caused by the ingestion of DDT. When one looks up the reference, however, it does not offer much support. DeWitt fed quail DDT at doses roughly 3,000 times greater than any humans were ever exposed to in their diets and found that 80 percent of the quails' eggs hatched. The hatch rate in the control group, which was not fed DDT, was 84 percent. That's hardly a spectacular difference, especially when the huge dose is considered. Carson was also silent about the fact that in the same study, eggs from pheasants fed DDT had a higher hatch rate than control groups. Her claims about declining bird populations were also suspect. The annual Christmas Bird Count of the Audubon Society, which itself claimed that DDT was harming birds, showed higher numbers in 1960 than in 1940—that is, higher numbers in the years in which DDT was most heavily used. Eagles, ospreys, and peregrine falcons had hit their population lows before DDT was introduced. Studies of eggs in museums have shown egg thinning half a century before DDT hit the market, perhaps due to acidification of the soil, which makes calcium less available. The culprit? Likely acid rain, the result of burgeoning industries.

What about the possibility that DDT is harmful to humans? It certainly presents no acute toxicity. You can eat the stuff. I wouldn't recommend it as a dietary staple, but Dr. J. Gordon Edwards of San José State University has no problem with it. Edwards is recognized as a world expert on the toxicity of insecticides and is a vocal critic of DDT's opponents. He would often begin lectures on DDT by eating a spoonful of it. After retiring, Edwards still climbed mountains well into his eighties. Human volunteers have eaten 35 milligrams of DDT for two years with no consequences. Of course, of greater concern is the possibility of chronic effects. Can DDT cause cancer in the long run? Apparently not. Over eighty peer-reviewed scientific publications attest to this fact, many of them having followed workers with extensive DDT exposure. One of the latest studies examined the increased incidence of breast cancer on Long Island, New York, where spraying with DDT in the 1960s was extensive. National Cancer Institute researchers found no link between tissue samples of DDT and the disease. This news was comforting because DDT does have possible hormone-disrupting effects, and a connection to breast cancer had been hypothesized. Certainly, the presence of compounds that can have "gender-bending" effects in the environment is undesirable. But not as undesirable as people dying. And they are dying of malaria in droves. Why Ceylon stopped spraying DDT in 1964 is a point of contention, but what is clear is that by 1969 the country was again recording 2.5 million cases of malaria per year. In 1996, due to environmental pressure, South Africa stopped spraying DDT. A malaria epidemic ensued and was only curtailed when DDT application was resumed four years later. Today, there are roughly 300 to 500 million cases of malaria per year worldwide and 2.7 million deaths. The judicious use of DDT has the potential to make a huge impact. Yes, there are concerns about insects developing resistance; indeed, that was one of the reasons cited for banning DDT. Actually, farmers who overused DDT on cotton crops fostered the resistance—not people who used DDT to control mosquitoes. There is also clear evidence that DDT even deters mosquitoes that are resistant to its insecticidal properties. For less than $1.50, a house in Africa can be protected for an entire year. The world may not need to use DDT for agriculture any more, and certainly nobody is advocating misting city streets in North America with DDT for mosquito control (as was once the case), but banning its use totally does not seem to me scientifically justified. You, of course, may have a different view. But do take the time to investigate all of the research before coming to a conclusion about Rachel Carson's true legacy.

FEEDING THE HUNGRY

I have never really been hungry. Sure, there have been occasions when I could hardly wait to wolf down a slice of pizza or a serving of veggie goulash, and I'll admit to looking forward eagerly to the end of the traditional Yom Kippur fast. But frankly, it isn't very hard to fast for twenty-four hours after you've filled your stomach with matzo ball soup and roast chicken. And the hunger pangs aren't too bothersome when you know that a superb stuffed cabbage awaits, ready to fill your belly. I suspect that most people reading this book have never felt real hunger either. Sure, we know hunger exists. We've read that 5,000 Africans die every day from a simple lack of food, and millions of others have diets that are totally inadequate to maintain health. These numbers may shock us, they may linger in our minds for a few minutes, but then they are squeezed out by our "real concerns": How many times per week can we eat canned tuna? Should we peel our fruits and vegetables for fear of pesticide contamination? Should ice cream be labeled a genetically modified food if it contains an emulsifier derived from corn that has been genetically altered to protect itself from pests?

Dr. Florence Wambugu finds these worries curious. Sometimes she is downright angered by them. You see, Dr. Wambugu knows hunger. She has experienced it personally and sees it around her constantly. Recently I had a chance to interview this remarkable woman on CJAD radio in Montreal, Canada, and experience her passion about using biotechnology to feed the hungry. Dr. Wambugu puts issues into perspective very quickly. "You people in the developed world are certainly free to debate the merits of genetically modified foods," she quips, quickly adding, "but can we please eat first?" And if Florence Wambugu is allowed to pursue her efforts to improve crop yields in Africa through biotechnology, she may help make that question redundant.

As a child living in Kenya, Florence Wambugu never dreamed that one day *Forbes* magazine would label her as one of the fifteen people "most likely to change the world." But that is precisely what happened in 2002. And it was all made possible by a cow. More specifically, the only cow Florence's family owned. Young Florence worked in the fields with her mother trying to raise crops to feed a hungry family of ten children, but their efforts were often stymied by plant diseases and ravenous insects. The mixtures of ashes and soot they used to ward off pests just didn't work. But Florence's mind did. Even at that young age she was inquisitive and bent on improving crop yields. Her mother, seeing the dedication, did the unthinkable. She sold the family cow so that Florence could go to secondary school. At the time it was a highly unusual move for a parent in Kenya to make, but, as it turned out, it was a great investment.

Florence eventually graduated from the University of Nairobi and went on to earn a master's in plant pathology in the US and a doctorate in England. Her research had always focused on ways to improve the lot of the African farmer, and her efforts were brought to fruition when she received a fellowship from the US Department of Agriculture to work with Monsanto scientists in St. Louis, Missouri. Dr. Wambugu had always been interested in the sweet potato—the crop her family had grown. It is resistant to drought and is filling and nutritious. But it is also very susceptible to attack by worms and the feathery mottle virus. African yields of sweet potato are the lowest in the world (as is the case with many crops), partly because there is no winter freeze to kill off pests. Florence spent years trying to crossbreed heartier varieties of sweet potato with little success. Then, in St. Louis, she tried a different tack. Chrysanthemums are known to produce chemicals called "pyrethrins," which are among the most effective natural pesticides known. Florence envisioned taking the gene that codes for the production of these chemicals from a chrysanthemum and splicing it into the DNA of a sweet potato. After all, Monsanto scientists had worked out a number of such gene-splicing techniques.

Although inserting outside genes into plants is now a relatively routine procedure, it doesn't happen overnight. Florence Wambugu worked for years to isolate and insert the appropriate gene and then tested the modified sweet potato in an isolated green house for two years. Finally, she received approval from the Kenyan government to begin limited field trials. Unfortunately, the early results are not quite what had been hoped for, but the experiments will continue. And so far the fields have not been ravaged by anti-biotech activists who (on a full stomach) often rip up such crops, claiming that they have been inadequately tested. The activists have, however, launched massive publicity campaigns to "inform" Africans about the "risks" of genetically modified foods and have labeled Florence Wambugu a corporate lackey. They have obviously not met her. I have.

The allegations have had some remarkable effects, including scaring some farmers away from any kind of new technology. A good example is the tissue culture that Dr. Wambugu has helped to develop. It involves taking tissue from a healthy plant and growing it in a sterile environment before planting the resulting seedlings in the field. In the case of bananas, this procedure greatly lessens the plant's risk of attack by fungi and bacteria, increases yields, and reduces the need to clear virgin land for farming. Tissue culture has nothing to do with genetic modification, but some farmers don't want to buy the seedlings because activists have warned them about the dangers of biotechnology. Perhaps even more disturbing is the comment made by Zambia's high commissioner to the UK. When asked why his country, faced with widespread and overwhelming starvation, would not allow the US to donate genetically modified corn, a crop that has been an integral part of the American food supply for years, he replied, "The fact that the people are starving doesn't mean that we should allow them to eat what they don't know." Such a response stuns the mind. But it doesn't do much for African hunger.

SAY "CHEESE"

Cheese producers were cheesed off. People were just not eating enough veal. Slaughterhouses were running short of calf stomachs and the cheese industry was feeling the pinch. There was not enough rennet to meet the demands of turophiles (that is, cheese lovers; the Greek word *turo* means "cheese") around the world.

Rennet, you see, is critical to the cheese-making process. At least it is if you want to indulge in something that is a little more exciting than cottage cheese. Traditionally, rennet has been made by washing, drying, macerating, and brining the lining taken from the fourth stomach of calves. This process leads to a product that is a mixture of two enzymes, chymosin and bovine pepsin, both of which can coagulate milk and convert it into cheese. Why does the stomach lining of mammals contain these enzymes? Because they are needed for proper digestion. If milk did not coagulate to some extent in the stomach, it would flow through the digestive tract too quickly and its proteins would not be sufficiently broken down into absorbable amino acids.

Enzymes are specialized protein molecules that serve as biological catalysts. They make possible the myriad of chemical reactions that go on inside our bodies all the time. Specifically, chymosin and bovine pepsin are proteases, which means that they catalyze the breakdown of proteins, a task that is central to the milk coagulation process. Milk consists of about 87 percent water, 5 percent lactose, 3.5 percent fat, 3.5 percent protein, and 1 percent minerals. The protein content consists mostly of casein molecules, which are insoluble in water and aggregate into tiny spheres called "micelles." Since their density is comparable to the surrounding solution, micelles remain suspended. Actually, there are three kinds of casein molecules: alpha-, beta-, and kappa-casein. Within the micelle, the alpha- and beta-caseins are curled up like a ball of string and are held together by kappa-casein, which functions much like an elastic band. The job of chymosin is to break the band and allow the casein molecules to stretch out and form a long, tangled network of protein molecules that settles out of the solution. Fats and minerals get snared in this protein net and—presto! We have cheese!

Chymosin is the ideal enzyme for catalyzing this process. In an acidic environment it snips kappa-casein specifically, allowing the other caseins to unwind. In the stomach, cells that secrete hydrochloric acid create the acidic environment whereas in cheese making, a bacterial starter culture that converts lactose into lactic acid is added before the rennet. Bovine pepsin is not quite as suitable as chymosin because it has a more general protease activity, snipping caseins in a variety of ways. This enzymatic action weakens the protein network needed to trap fat and results in a lower yield of cheese. Furthermore, some of the protein fragments it produces have a bitter taste and subtly alter a cheese's flavor profile.

By the 1960s the shortage of rennet was becoming critical. The stomachs of older animals were pressed into service, but the resulting rennet was not really suitable. As an animal ages, chymosin production decreases and pepsin production increases.

So scientists had to step in and take the bull by the horns, as it were. Actually, instead of bull horns they grabbed chicken bones. In Canada, researchers at the University of Guelph discovered that rennet enzymes would bind very well to porous chicken bones and milk could be pumped through a matrix of these bones to start the curdling process. With this procedure, the same amount of rennet would go a lot further than if it were just dumped into vats of milk. While an interesting possibility, the method was never commercialized because a number of companies found that by using a technique called "anion exchange chromatography," they were able to separate pure chymosin from the stomach extracts of older animals. This technique made 100 percent pure chymosin available for the first time, but the process was complicated and not cheap. However, there was another way to compensate for the lack of calf rennet.

In the 1960s, researchers had discovered that certain fungi, *Mucor miehei* being a prime example, were capable of producing enzymes that would cleave proteins in much the same way as chymosin. This discovery meant that cheese could be produced without using any animal rennet at all. The breakthrough not only addressed the rennet shortage, but it also made possible the production of cheese that met the needs of vegetarians. Such cheese was also kosher because there was no mixing of milk and meat during production. But purists claimed the taste was not the same. They may well have been right. Fungal enzymes have greater proteolytic, or protein-breaking, activity than chymosin and can give rise to "off" flavors.

Then genetic engineering entered the picture and essentially solved the chymosin problem. The bit of DNA, the gene, that gives the instructions for the formation of chymosin was isolated from calf cells and was copied, or "cloned." It was then successfully inserted into the genetic machinery of certain bacteria (*E. coli*), yeasts (*Kluyveromyces lactis*), and fungi (*Aspergillus niger*), which dutifully churned out pure chymosin. Approved in 1990 by the US Food and Drug Administration, chymosin became the first product of genetic engineering in our food supply. It is 100 percent identical to the chymosin found in calf stomach, but because it does not come from an animal it is acceptable to consumers who do not want meat products in their cheese.

Extraordinary precautions were taken before chymosin, made by recombinant DNA technology, was marketed. Regulators ensured that no toxins of any kind had been introduced and that no live recombinant organisms were present. Indeed, the product contained nothing but pure chymosin. Cheese made with it is completely indistinguishable from that produced with animal rennet. In any case, chymosin itself is degraded during cheese making and none is left in the finished product. Today in North America, over 80 percent of all cheese is made with chymosin produced by recombinant DNA technology. Cheesemakers no longer have to worry about a shortage of calf stomachs, and turophiles can satisfy their critical taste buds. Thanks to biotechnology, they can say "Cheese" and smile.

THE TEFLON SCARE

I knew something was going on when I listened to my messages one morning. The first caller wanted to know if it was safe to keep using Teflon dental floss and the second inquired about the best way to dispose of her Teflon cookware. The third wanted to know if it was safe to keep wearing a Teflon-coated hat. It didn't take me long to find out that the scare had been triggered by a segment on the ABC news program *20/20*, which had addressed some concerns about Teflon the previous night. This issue follows on the heels of the closely related "fabric protector" story. Let's start with that one.

Back in 1952, Patsy Sherman, a young chemist at the 3M Company, was assigned the problem of finding a material that was flexible and could stand up to the corrosive nature of jet aircraft fuels. Gaskets and hoses commonly deteriorated and had to be frequently replaced. Sherman was familiar with the chemical-resistant properties of fluoropolymers such as Teflon and began to experiment with similar substances. The research seemed to be going nowhere until one day Sherman's assistant accidentally spilled a few drops of a novel compound on her new tennis shoe. She became frustrated because neither water, alcohol, nor any other solvents were able to remove the stain. Sherman was quite taken by the material's repellant properties and shifted the focus of her research. By 1956, with the help of fellow 3M chemist Sam Smith, Scotchgard Protector made a triumphant entry into the marketplace as a virtually magical substance that repelled water and stains from clothes, carpets, and furniture fabrics.

Scotchgard was an immensely successful product line, with active ingredients being manufactured by the millions of pounds annually. Then, all of a sudden, in May 2000, 3M made a startling announcement: it was phasing out the manufacture of perfluorooctanyl sulfonate (PFOS), the key chemical used to produce Scotchgard products. The company's chemists had found that Scotchgard can degrade to release PFOS, which was turning out to be more persistent in the environment than they had previously believed it to be. It had been detected, albeit at very low levels, in the blood of seals, dolphins, minks, bald eagles, and, most important, humans. For years the company had been monitoring blood levels in employees working with the chemical and became concerned that uptake of PFOS was exceeding the body's ability to excrete it. But the problem came to a head when PFOS was found in samples from blood banks that were to be used as control samples for the workers' blood. It soon became clear that all humans had some PFOS in their blood. How was it getting there?

Studies have shown that about one-third of a sprayed product is lost into the air, ready to be inhaled by people. Given that discovery, and the fact that huge amounts of the repellant chemicals were used in products ranging from fast-food packaging and linens to tents and upholstery, it comes as no surprise that remnants show up in human blood. Of course, just because a chemical is present in the blood it does not mean that it presents a danger. In the case of PFOS, however, there was an indication from rat and primate studies that excessive exposure may be of concern. No human health problems have ever been linked to PFOS and 3M maintains that its withdrawal of the chemical was based on environmental concerns. Perhaps. Or maybe the company saw the writing on the wall and decided to take action before being forced to do so by the US Environmental Protection Agency (EPA). In any case, 3M has been successful in developing alternate formulations for many, but not all, uses of Scotchgard. The new key ingredient is a smaller molecule, perfluorobutyl sulfonate, which, we are told, is nontoxic and nonpersistent.

When 3M phased out PFOS, it also stopped production of perfluorooctanoic acid (PFOA), which it sold to other companies for use in the production of Teflon. There is no viable substitute for this compound in the manufacture of Teflon, and Dupont now produces large volumes of it. Like PFOS, it too has been found throughout the environment. In this case, the source is not obvious because finished Teflon products do not contain any PFOA. One possibility is that another type of stain-repellant material made of short-chain fluorinated polymers, called "telomers," breaks down to release PFOA.

Most people had never heard of PFOA until *20/20* focused on it. Highlighting the case of an unfortunate young man born with one nostril and a deformed eye, the piece noted that his mother had worked in Teflon production while pregnant and inferred that exposure to PFOA was responsible for his defects. Birth defects are not uncommon and it is unscientific to make such a link without more evidence. The program also went on to describe how heating Teflon to temperatures above 290°C (554°F) can cause the release of fumes that are toxic to birds and can cause a reversible "polymer fume fever" in humans. But these facts in no way mean that Teflon dental floss, hats, or cookware used properly present a risk to consumers. The persistence of PFOA is an issue, and the EPA is looking into it. But the agency's investigation has nothing to do with using the pots and pans in your kitchen. Just use Teflon cookware as it is meant to be used and don't fry foods at extreme temperatures. In any case, if you are heating foods to 290°C (554°F), you had better worry more about the toxic compounds formed by the heating process than the ones released by Teflon. This type of cookware

actually lets you cook foods with less fat so that the end product is healthier. On that note, I think it's time to have some
lunch: stirfried vegetables—cooked in a Teflon pan, of course.
After lunch I'll use some Teflon dental floss. And if I had a
Teflon-coated hat, I'd happily sport it.

Weeding Out the Pesticide Myths

Pesticides have one indisputable effect: they cause emotions to
boil over. That's just what happened when a group of golfers
noticed that a chemical sprayer was out on the course as they
were completing their round. By the time they got into the
clubhouse, several were complaining of headaches, rashes, and
general malaise and angrily approached the superintendent to
protest what they believed was an irresponsible activity. The
golfers linked their symptoms with the chemicals being sprayed
on the grounds because they were convinced that the use of
pesticides is inherently unsafe. Were they right?

Asking if it is safe to use pesticides is like asking if it is safe
to take medications. The answer can be either "Yes" or "No"
depending on which medication is taken in what dose, and how,
by whom, and for what reason it is taken. Salt, vitamin B-6,
vitamin A, and caffeine, on a weight-for-weight basis, are more
toxic than many pesticides. Basically, instead of classifying substances as "safe" or "dangerous," it is far more appropriate to
think in terms of using substances in a safe or dangerous fashion.
Two aspirin tablets can make a headache go away, but a handful
of them can kill. Unfortunately, in rare cases, even two tablets
can cause side effects. So it is with pesticides. While there are safe
ways to use these chemicals, there can be no universal guarantee
of safety. Remember, pesticides are designed to kill their targets,
be they insects, weeds, or fungi. The best we can do is evaluate

the risk-benefit ratio of each substance and make appropriate judgments.

In Canada, such judgments are made by Health Canada's Pest Management Regulatory Agency (PMRA). Before a pesticide can be registered for use, the agency's toxicologists, physicians, chemists, and agronomists have to be convinced that the substance in question can effectively handle the problem it was designed to solve and that its risk profile is acceptable. A registration is a long and involved process requiring acute, short-term, and lifelong toxicology studies in animals as well as studies of carcinogenicity and possible damage to the nervous system. Proof of absence of birth defects is required. A pesticide's effects on hormone levels has to be studied in at least two species along with its effects on nontarget species. All routes of exposure are assessed—ingestion, inhalation, and skin contact—and cumulative effects are studied. PMRA also requires field testing for environmental effects before a pesticide is approved.

Based on all the data, PMRA assesses a pesticide's risk, taking into account exposure of children, pregnant women, seniors, pesticide applicators, and agricultural workers. The potential level of exposure can be no more than 1/100 of the dose that showed no effect in animals. Even once a pesticide is registered it undergoes a continuous reevaluation system that includes the "inert" ingredients used in its formulations. Risk assessments are refined in accordance with new research findings. All ways of reducing pesticide risk are examined, with great emphasis on integrated pest management, or IPM, which is aimed at reducing reliance on pesticides as the sole approach to pest management. IPM is geared toward taking action only when numbers of pests warrant it and uses a mix of biological, physical, and chemical techniques. Furthermore, PMRA has inspectors across the country to monitor the use of pesticides.

It is hard to imagine what more could be done to ensure that a pesticide has an acceptable risk-benefit ratio. But can even such a rigorous system ensure that we will experience no consequences from the use of pesticides? Absolutely not. There may be subtle effects in humans that show up only after years of exposure. These can be revealed only by long-term studies, not by anecdotal evidence. Pesticides cannot be linked to cancer on the basis of a heart-wrenching story appearing in the media that describes how a child who had repeatedly felt ill after exposure to lawn sprays was later diagnosed with cancer. Long-term epidemiological studies are required to make that kind of link. A number of such investigations have been carried out.

Workers in the agricultural chemical production industries—who would be expected to have the highest exposures to pesticides—do not show any unusual disease patterns, but the number of subjects in these studies is small. A widely reported study of farmers who sprayed their fields with pesticides showed a weak link between acres sprayed and various cancers, but overall the group of farmers studied had fewer cancer cases than the general population. A frequently cited American study seemed to indicate a link between non-Hodgkin's lymphoma and acres sprayed with the herbicide 2, 4-D, a chemical that is also used in home lawn care. But a long-term study of workers who manufactured 2, 4-D and had huge exposures over many years showed no increase in the incidence of cancer.

One of the developing concerns about the use of insecticides and herbicides is a possible effect on the immune system. Laboratory evidence indicates impaired activity of immune cells after exposure, and at least one study has shown increased respiratory infection in teenagers in villages where pesticide use is the heaviest. There is also the possibility of neurobehavioral effects. In a Mexican study, children living in areas where pesticides were extensively used performed more poorly on coordination and memory tests. But the conditions in these studies are very different from those involved when a dilute solution of 2, 4-D is occasionally used on a lawn by trained applicators. In fact, studies at Guelph University in Canada have shown that even walking barefoot on a lawn an hour after treatment leads to no detectable levels of 2, 4-D in the urine. But it is possible that home gardeners who purchase pesticides and use them improperly can put themselves and others at risk.

It would be great if we could get away from using pesticides. No exposure to pesticides means no exposure to their risks. At home, we can manage to avoid these chemicals. After all, a few dandelions on the lawn are not life threatening. In fact, quite the opposite is true: they can be made into a nutritious salad! But we cannot feed a global population of 6 billion without the appropriate use of agricultural chemicals. So we do have to put up with risks, both real and imagined, because on a global scale they are outweighed by the benefits. And just what was that dastardly chemical being sprayed on the golf course, the one that caused such severe reactions in the golfers? Good old H_2O! Fear itself can sometimes be hazardous.

Paralysis through Analysis

Measuring the size of an alligator's penis is not an easy job. It entails cruising around in a boat at night, scanning the waters of a Florida lake with a flashlight, and looking for the red reflection from the gator's eyes. After locating an animal, you approach it silently—hoping it's a male—grab it behind the head, and hoist it into the boat. Then you get out the calipers and measure the creature's manhood. Since I suspect the alligator is not a willing participant in this process, he'll probably be in a pretty foul mood, a mood that is not likely to improve when he is jabbed a few times with a needle for a blood sample. But this is the kind of job you do if you are a wildlife biologist interested in the welfare of alligators and, possibly by extension, the health of people.

Let's go back for a moment to 1985 when Dr. Lou Guillette, then a young biology professor at the University of Florida, became interested in the environmental impact of ranchers harvesting alligator eggs from lakes. Lake Apopka was an appropriately large lake for study, but much to Guillette's surprise, alligator eggs in this lake were in short supply. And the eggs that he did find were of poor quality and less likely to hatch than eggs from other lakes. Furthermore, those that did hatch yielded alligators that showed a variety of abnormalities. The males had very low testosterone levels and the females had higher than normal estrogen levels. Somehow, the alligators in Lake Apopka had been "feminized." Dr. Guillette wondered if the problem was widespread. To answer that question he embarked on his penis project, which resulted in the subject of "endocrine disrupters" being thrust into the spotlight.

By 1995 Guillette had found that the penises of alligators in Lake Apopka were some 25 percent shorter than those of alligators caught in a similar lake. That finding was curious enough, but even more puzzling was the observation that in the healthy alligators, unlike in the Apopka animals, testosterone levels were clearly related to penis size. Even if the Apopka alligators had normal testosterone levels, their organs were length challenged. Somehow, the normal effects of testosterone were being blocked. Could there be a chemical in the water that was disrupting the activity of the alligators' hormones?

As early as 1938 scientists were aware of the possibility that certain chemicals, when introduced into the body, had hormone-like activity. Edward Charles Dodds had synthesized diethyl-stilbestrol (DES) in the laboratory and found that it mimicked the action of estrogen. At the time, physicians believed that miscarriages were linked to low levels of estrogen and began giving DES to women who had a history of problem pregnancies. Not only was DES not the solution, but it was also later linked to a rare form of vaginal cancer in the daughters of women who had taken the drug during the first three months of pregnancy. In the 1950s researchers found that the insecticide DDT had an estrogen-like action on roosters, stunting the growth of their combs and testes. This finding did not garner much attention, however, because DDT had performed so spectacularly in controlling insects and reducing the incidence of diseases such as malaria. It was a chemical that was produced on a massive scale, including at the Tower Chemical Company on the shores of Lake Apopka.

When Lou Guillette followed up on his hypothesis that something in the lake was affecting alligator hormones, he discovered that in 1980 a waste pond at the Tower Company had overflowed and DDT had been discharged into the water. There was the smoking gun! But Guillette still had to match the bullet to the weapon. Analysis of the alligators' tissues showed the presence of DDT as well as its breakdown product, DDE. Of course, the mere presence of these chemicals did not prove that they

were causing the effects found in the animals. Guillette then took some healthy alligator eggs from a different lake and treated them with a mix of DDT and DDE at concentrations found in the Apopka gators. As he had suspected, the treated alligators matured into adults that were "short in stature" where it counted. Guillette's findings unleashed a tabloid bonanza. Most men didn't care about the alligators' shortcomings, but they sure were interested in the subject of endocrine disrupters when they discovered that DDT and DDE were ubiquitous in the environment and present, albeit in tiny amounts, in just about everyone's body. The issue affected women as well, since in theory, estrogen-like substances could be linked to breast cancer.

While stories about alligators' privates were arousing the public's attention, an experiment at Tufts University in Boston was raising scientists' eyebrows. Test tubes may not be as romantic as penises, but the work of Drs. Ana Soto and Carlos Sonnenschein would turn out to be seminal in the study of endocrine disrupters. These researchers were investigating the effect of estrogen on breast cancer–cell proliferation when they noted something unusual. In some of their test tubes the cells were proliferating, even though no estrogen had been added! They remembered receiving a new shipment of tubes and wondered if there had been a change in their formulation; perhaps some estrogenic component was leaching out. The manufacturer was unhelpful. While acknowledging that there had indeed been a change in the product's chemical makeup, "proprietary reasons" prevented its disclosure.

A little chemical detective work identified the mysterious ingredient as nonylphenol, a compound added to the polystyrene tubes to make them less brittle. Was the nonylphenol the estrogenic component? Laboratory experiments indicated that it was, and confirmation came when it was injected into female rats whose ovaries had been removed. Nonylphenol caused the same kind of changes in the lining of the uterus that would be caused by estrogen produced by the ovaries. Nonylphenol was definitely estrogenic! And as Soto and Sonnenschein would soon discover, the presence of nonylphenol wasn't limited to some lab equipment. Nonylphenol was an ingredient in various plastics as well as some contraceptive creams. Furthermore, it was also a breakdown product of certain detergents. In other words, hormone disrupters lurked in everyday items.

People may not get too excited about alligator micropenises, but when they learn that there may be hormone-like chemicals in their food supply, they get nervous. So do some researchers. One of those researchers is Dr. Fred vom Saal, a reproduction biologist at the University of Missouri, who I'm sure would not approve of some of my daily activities. Like eating cheese that has been wrapped in plastic. Or putting berries from a plastic box on top of my morning bowl of steel-cut oats. Or downing yogurt from a plastic container, peas from a can, or water from one of those 5-gallon carboys perched atop a cooler. Or showering twice; once in the morning, and once after working out at the gym.

These may seem pretty benign activities to most people, but Dr. vom Saal doesn't share that view. He is highly suspicious of chemicals that can leach out of plastics and goes so far as to avoid all canned foods because of concern about the protective plastic layer that lines the insides of the cans. He worries about endocrine disrupters, the hormone-like compounds made famous by the shrunken manhood of those Florida alligators. Vom Saal says that these endocrine disrupters even show up in our shower water—probably the result of leaching from plastics in landfills—ready to be absorbed through the skin.

Professor vom Saal is no academic slouch. He earned a Ph.D. from Rutgers University in New Jersey, spent a couple of years in the Peace Corps teaching biology, and carried out postdoctoral research at the University of Texas before taking up his post at the University of Missouri. But when Dow Chemical, Shell Oil, or General Electric makes up a guest list for a company picnic, Dr. vom Saal's name is not on it. That's because, as he himself says, in 1995 he "stepped on an elephant's toe." Until 1990, Dr. vom Saal had been studying the effects of fluctuations of naturally occurring hormones on mice fetuses. In response to his publications, he received a phone call from Dr. Theo Colborn, a scientist at the World Wildlife Fund in Washington, DC, who had detected gender ambiguities in animals in the wild and suspected that some hormone-mimicking chemicals had found their way into the environment. Dr. vom Saal then began to test substances off the laboratory shelf to see if any would indeed act as hormones. He discovered that a compound called "bisphenol A" did just that. He didn't realize at the time that this chemical was made industrially to the tune of some 2 billion pounds a year, and that it was used in the manufacture of substances ranging from baby bottles and dental sealants to the linings of food cans.

In an experiment that changed his professional life, Dr. vom Saal treated mice with fantastically small amounts of bisphenol A and noticed the result was enlarged prostate glands in the mice fetuses. He presented his results at a conference and obviously triggered the interest of bisphenol A manufacturers. And so began an era of accusations and counteraccusations. The plastics industry maintained that effects at such low levels ran counter to accepted principles of toxicology; bisphenol A had been thoroughly tested in animals at doses much higher than those administered by vom Saal without adverse effects. How, then, could much smaller doses produce the claimed results? Dr. vom Saal retorted that there were flaws in the concept that adverse effects are directly proportional to the dose. He argued that in the case of hormone-like chemicals, a tiny dose in the fetus could cause effects unseen in adults given even higher doses. In response to this claim, industry researchers tried to reproduce vom Saal's original experiments without any success and raised questions about the credibility of the work. In turn, von Saal questioned the reliability of the industry's studies. Now a chance finding at Case Western Reserve University in Ohio has brought the allegations that have been flying back and forth in the scientific literature to the attention of the public.

Dr. Patricia Hunt was studying reproduction in mice when she noted a higher than normal increase in abnormalities in developing egg cells. Weeks of research did not reveal a cause. But then Dr. Hunt noticed that the clear plastic cages in which the animals were kept "looked a little worse for wear." It seems that the technician in charge of the animals had used a new alkaline detergent to clean the cages, which resulted in the leaching out of bisphenol A. Dr. Hunt then deliberately exposed her mice to low doses of bisphenol A and found the same abnormalities. This scientific finding is certainly noteworthy because such chromosomal changes, were they to be found in humans, could cause Down's syndrome and miscarriage. But—and this is a very important but—the study did not show that these changes could occur in humans, and the researchers did not examine any health effects on the animals. Dr. Hunt herself says she has no idea about the relevance of this finding with regard to humans.

Bisphenol A producers say that they have a pretty good idea of what the relevance is: practically none. They cite published studies in which two generations of animals have been treated with both higher and lower levels of bisphenol A than the dose used in Hunt's study, with no effect. They also point out that polycarbonate baby bottles, which have been attacked as sources of bisphenol A, have been thoroughly studied, including after repeated filling and heating. No bisphenol A at the remarkably low detection limit of 5 parts per billion was found. This result is of no comfort to Dr. vom Saal, who maintains that levels defying laboratory detection can still cause an effect in animals.

Others researchers, such as world-renowned toxicologist Dr. Stephen Safe of Texas A&M University, don't buy this argument. While admitting that endocrine disruption is certainly a concern, Dr. Safe points out that synthetic-hormone mimics contribute less than 1/1,000 of 1 percent of the amount of estrogenic compounds that people consume in their diets. Natural-hormone mimics in foods, which are found in a variety of grains, fruits, and vegetables, are far more prevalent. Furthermore, there is no scientific evidence to suggest that the body somehow handles these natural-hormone mimics differently from synthetics. To add further confusion, it seems that sometimes bisphenol A may even be beneficial! A research group led by Debby Walser-Kuntz at Carleton College in Minnesota found that in lupus-prone mice, the disease took longer to develop and was less aggressive when the animals were treated with doses of bisphenol A comparable to what humans may be exposed to in the environment.

And on it goes. So where does the duel between scientists leave the public? Confused. On the one hand, some researchers suggest that hormone mimics may be responsible for an increase in prostate and breast cancer in humans. On the other hand, many scientists maintain there is no evidence for this claim.

What do I make of the conflicting theories? If we start worrying about chemicals at concentrations of a few parts per billion, we'll be subject to "paralysis through analysis" and driven to despair. We are exposed to hundreds of thousands of compounds, both natural and synthetic, at such levels. It is impossible to know the consequences of these exposures. So I think I'll concentrate on things that we know really matter, such as eating more fruits and vegetables (wrapped in plastic or not) and exercising. And I'll continue to take showers after my workouts. But if I ever end up with mice for pets, I won't use caustic detergents to clean their polycarbonate cages.

Bottled or Tap?

Jackie Chan's movie *The Tuxedo* may not be a candidate for an Oscar, but it does have a great opening sequence. The camera draws back from a cascading mountain stream to reveal a deer relieving itself into the pristine water. We then follow the water on its journey through rivers and pipes until it ends up in a plant that bottles water. The water is treated along the way, of course, but it is evident that the source of the bottled water may be no different than the source of the stuff we get virtually for free from the tap. So what's bottled water doing in a Jackie Chan movie anyway?

Well, the action predictably features a dastardly criminal bent on world domination. And a requisite mad scientist. Together they concoct a scheme to contaminate all the water in the world with a bacterium that will almost instantly dehydrate anyone who drinks from a municipal water supply, turning them to dust. But the villain is not a vacuum cleaner manufacturer; rather, he just happens to be a bottled water magnate who will own the only safe water supply in the world after the bacteria are unleashed, allowing him to achieve world domination. It's never quite clear what this guy will do with the world he dominates. What is clear, though, is that scriptwriters attempt to make their plots current by focusing on issues in the public eye. And everyone is talking about bottled water—not to mention drinking it. The familiar plastic containers are ubiquitous today; people sip from them constantly, apparently afraid that if they miss a dose they will drop from dehydration.

We are continually exhorted to follow the eight-by-eight rule: drink eight 8-ounce glasses of water per day to keep hydrated and to "flush toxins from the system." Where does this advice come from? As far as I can tell, it can be traced back to 1945 when US National Academy of Sciences (NAS) recommended that people consume roughly 1 milliliter of water for every calorie taken in. Accordingly, a 2,000-calorie intake would require about 2 liters of water, the equivalent of about eight glasses. How the NAS came to this conclusion is a matter of some mystery, and the eight-glasses-per-day guideline has never been scientifically validated. It has, however, become "fact" just by the process of repetition. I should point out that the NAS never suggested that we go out of our way to drink eight glasses every day. The recommendation clearly stated that most of the water we need is contained in the foods and beverages that we normally consume.

There is no scientific justification for forcing ourselves to drink eight glasses of water per day if we don't thirst for it. Our kidneys require about a half liter of water per day to excrete the waste products of metabolism, and we lose another half liter or so as we sweat, breathe, and eliminate feces. That makes

for a liter; a greater intake dilutes the urine but doesn't result in more toxins being excreted. "Flushing toxins out of the system" is usually a catchphrase used by unscrupulous marketers of various drinks and supplements who try to capitalize on peoples' fears. In some cases—people suffering from kidney problems or recurrent urinary tract infections, for example—advice to drink more water is legitimate. And intense physical activity obviously requires the replacement of lost water.

According to Dr. John Leiper, a researcher at the University of Aberdeen in Scotland, bedroom gymnastics may fall into this last category. He explains that British couples often develop headaches or become lethargic after romantic jaunts because they don't replenish the lost water. Apparently the French don't have this problem. They drink five times more water than the Brits and partake in carnal pleasures far more frequently. Sweating, flushing, and panting, Leiper says, lead to dehydration. A half hour of lovemaking, he claims, can be as strenuous as a 3-mile run at the gym. Now, I know all about that. I mean, of course, the 3-mile run. That just happens to be the distance I shoot for most every day, and it sure does require water replenishment. But it doesn't require any special kind of water, no matter what advertisers would have us believe.

The talk about "pure" bottled waters that have been treated to "alter the molecular structure" of the contents for efficient hydration is pure bunk—even if it comes from a physician, as in one advertisement I came across. It is really hard to understand how someone who has gone through medical school can come up with this gem of an endorsement:

> They [the manufacturers of the physician's pet product] have developed a physics process to convert purified water containing mostly large molecular clusters into water having a high concentration of small clusters. These are able to pass freely in and out of cell membranes, bringing nutrients and removing wastes. I have found Penta water to be especially beneficial for my patients with fibromyalgia, headaches, arthritis, as well as most degenerative diseases. I usually recommend two 16-ounce bottles a day.

I recommend that the good doctor take a course in chemistry and another in medical ethics.

Certainly, the bottom line here is not that water intake is not important. It is. But we don't have to neurotically attach our lips to a bottle all day to meet the mythical eight-glass agenda. A water intake equivalent to five glasses per day, though, does have some scientific justification. A study at Loma Linda University in California followed the health status of some 8,000 women and 12,000 men who had completed dietary surveys in 1976. Statistically, the 246 who eventually died from heart disease were more likely to have consumed less than five 8-ounce glasses of water per day. The researchers offered the rationale that high water intake may decrease blood "thickness" and lower the risk of the formation of blood clots, which can trigger a heart attack. As usual, one has to be careful about jumping to conclusions based on any one study. After all, how likely is it that the subjects maintained their 1976 dietary patterns? The press, however, welcomed the findings with unjustified headlines such as "Drinking more than 5 glasses of water can reduce the risk of heart disease by 40%!" I certainly have nothing against drinking that much water daily. Whether that water should be bottled or from the tap is an interesting issue.

Bottled waters have a long history. When Michelangelo was tormented by kidney stones, he found relief from drinking water every morning and night from a spring at Fiuggi, located about 40 miles from Rome. Peter the Great of Russia suffered from chronic indigestion, which was apparently eased by gulping twenty-one glasses of Belgian Bru water every day. George Washington had some constipation concerns that were alleviated by drinking Saratoga Springs water, and when President Eisenhower suffered a heart attack, his physicians advised him to drink Mountain Valley bottled water. The health properties of these waters were ascribed to their mineral content. But that was then. Now, people drink bottled waters not because of what is in them, but because of what is not. With all the talk about our water being polluted with mercury, lead, and various industrial chemicals, consumers are concerned that municipal treatment may be inadequate. They also worry about chlorination, which may take care of bacteria but can lead to the formation of undesirable trihalomethanes (THMs).

The bottled water industry has jumped on these fears and launched aggressive advertising campaigns to implant in the public's mind images of purity and health linked to unspoiled mountain springs. Swedish Ramlösa advertises that its water is "unharmed by man," and the Ice Mountain label features snowy peaks—an oddity considering the water comes from springs in Texas and Michigan. Perrier informs consumers that while "man makes love, makes war and makes fire, only the earth can make Perrier." Evian ads feature sweaty, sinewy, perfectly sculpted young bodies being refilled with the popular French water. The advertising obviously works. Raquel Welch washes her hair with Evian and Michael Jackson is said to bathe in it. Not the greatest testimonial, one would think. (Incidentally, there is no truth to the urban legend that the name *Evian* was chosen because it spells *naïve* backward.)

While some of the advertising may be questionable, there is no question that the bottled water industry is booming. Consumption has tripled in the last ten years, and water is poised to surpass beer, milk, and coffee in sales to rank behind only soft drinks in the highly competitive beverage industry. Even the big boys have gotten into the game. Pepsi produces Aquafina, whose label is adorned with pictures of mountains and ice, and Coke has come out with the curiously named Dasani. Apparently consumer testing showed that the name is relaxing and suggests "pureness and replenishment." Neither company promotes the fact that its source is the municipal water supply. Admittedly, each does subject the water to further purification, and in the case of Dasani, magnesium sulphate and potassium chloride are added to enhance flavor.

Once the images of melting glaciers and nubile maidens are stripped away, one fundamental question remains: Are bottled waters superior in quality to the liquid that flows for free from the tap? By and large, if we go by chemical analysis, the answer is "Yes." Whether the difference is of any practical significance is an entirely different question, one that is almost impossible to answer. Potential tap water contaminants are very carefully monitored, and drinking water supplies must adhere to stringent government regulations. Does the elimination of a few parts per million of THMs or trace amounts of lead make a difference over a lifetime? Short of designing a study that would follow two groups of people who drink either bottled water or tap water for decades while keeping everything else in their lives the same, we will never know. We do know, however, that bottled water often tastes better because it contains no chlorine residues. But to avoid the chlorine taste and to reduce levels of lead and other contaminants, cheaper alternatives are available. Tap-mounted filters such as Brita or PUR do an excellent job.

Furthermore, not all bottled waters are created equal. When the US Natural Resources Defense Council, a prominent environmental organization, analyzed over 1,000 bottles from 103 brands, it found that some had bacterial counts far higher than tap water, although still not in excess of safety limits. About one-third of the samples contained various synthetic organic compounds as well as arsenic—precisely the type of contaminants people fear may be present in tap water. Bacterial contamination of bottled waters is not a big issue, although there may be a concern if bottles are refilled with tap water without adequate rinsing. Remember that you may be introducing bacteria from your mouth into a bottle every time you drink from it and these may flourish if the bottle is not cleaned.

What about the practice of refilling these bottles? The clear plastic ones are made of polyethylene terephthalate and contain no plasticizers that can leach out (except perhaps minute amounts that are used in the polyvinyl chloride caps). Over the long term the plastic may degrade, resulting in water with a slight fruity flavor. This is no big deal, but the industry (for obvious reasons) does not suggest refilling the "one-time use" bottles. The bigger 4-liter opaque bottles are made of high-density polyethylene and may sometimes impart a bit of "melted plastic" taste to the water. The large 5-gallon clear plastic containers are made of polycarbonate and do not cause off flavors, but there is some concern that they leach small amounts of bisphenol A into the water. This substance is an animal carcinogen and also has hormonal effects, but once again the amounts released are in the parts-per-billion range. Nobody knows the long-term consequences of ingesting such tiny amounts of bisphenol A.

While there certainly is no great health risk in consuming water from plastic bottles, there is an environmental issue. The plastic is made from petroleum, a nonrenewable resource. Bottles can be recycled, but only into lower-grade plastics suitable for items such as park benches. Also, a great deal of energy is consumed by the assembly lines that fill the bottles and, of course, transporting the bottles by truck uses still more energy.

So what's the bottom line here? If you don't like the taste of tap water, use a filter. Bottled water is good too—especially for those who sell it. What about those stories about Michelangelo flushing his kidney stones out with water from the springs at Fiuggi? I suspect any water would have done just as well. I could go on about bottled water but I just lost my taste for the subject. I came come across a reference for an Australian "organic water" that is sourced "from organic rock springs in Victoria's Snow Mountains." Organic water? Now that's tough to swallow.

Waxing Lyrical about Fruit Wax

Apparently Snow White's wicked stepmother knew something about chemistry and toxicology. She brewed up a remarkable poison to put the anemic young lady into a state of suspended animation. But how to get her to ingest it? Who, thought the evil queen, could resist a bright, shiny apple? Certainly not Snow White. And you know the rest of the story.

Well, according to some current fear mongering, you don't need a depraved queen to taint apples with poison; you just need some apple processors who apply wax to the fruit. The wax may contain a compound called morpholine, which some believe presents a risk to humans. Why? Because morpholine, under certain conditions, can be converted into nitrosomorpholine, a compound known to cause cancer in rodents. Perhaps a little reality check is in order.

Most fruits and vegetables are naturally covered with a thin layer of wax. This barrier prevents moisture loss and makes it harder for fungi to get a foothold in the fruit. You can pick an apple straight from the tree and note the waxy coating—it is not synthetic. When processors wash apples in order to remove dirt, microbes, and pesticide residues, much of the protective waxy layer is lost. As a result, the fruit loses moisture more readily, is more susceptible to attack by fungi, and looks less appealing. To counter this problem, a variety of waxes that can readily replace the fruit's natural wax have been developed. Almost all these waxes are derived from one of the following sources: beeswax, carnauba wax, candelilla wax, shellac, or oxidized polyethylene. Beeswax, of course, is processed from honeycombs; carnauba and candelilla waxes come from plant leaves; and shellac derives from the resin that Indian "lac" bugs secrete to protect their eggs. Polyethylene is a synthetic plastic that, upon reaction with chemicals such as potassium permanganate, is converted into a waxy substance that can adhere to fruit.

Sometimes, under conditions of high humidity, the waxy coating cracks and the fruit's surface takes on a milky appearance. While cosmetically unattractive, this state does not present any added risk. You will note that it is almost impossible to wash off the wax, as it is not soluble in water. If you are afflicted with "waxophobia," peeling the fruit is the only answer. Some people routinely peel fruit because they fear the wax seals in pesticide and fungicide residues; others advise against peeling because they fear the loss of nutrients contained in the peel. Actually, peeling fruit in order to get rid of microbes is probably a more realistic scenario than peeling it to eliminate pesticide residues. As far as loss of nutrients goes, no need to worry: if you are eating the recommended five to ten servings of fruit per day, the nutrients lost to peeling are insignificant.

The amount of wax applied to any individual piece of fruit is extremely small. Only about 0.1 percent of the fruit's final weight is due to the wax. In order to affix such a thin coating, the wax is mixed with a solvent, usually a combination of water and alcohol, and is sprayed onto the fruit. A thin waxy layer is left behind as the solvent evaporates. To ensure that the wax is uniformly dispersed in the solvent, a variety of processing chemicals are used. One of these substances is morpholine, an emulsifier that helps distribute the wax evenly. This is the chemical that has been targeted as a cancer risk in some press reports.

So how much morpholine is in the wax? A typical range is 3 to 4 percent by weight of the wax solution that is applied. Morpholine is volatile, meaning that some of it evaporates during application. The amount left on the fruit is in the range of millionths of a gram. How much of the residual morpholine is absorbed into the body? This measurement is almost impossible to determine, but given the wax is indigestible (meaning that most of it comes out the other end) and the morpholine is embedded in the wax, the amount is likely trivial. Furthermore, morpholine itself (chemically an "amine," an organic compound derived from ammonia) is not the problem. To pose a concern, it has to undergo "nitrosation" to form nitrosomorpholine, which indeed is a carcinogen. Theoretically, this process can happen in the body, but the amount of nitrosomorpholine formed would be extremely small. And why worry only about nitrosating morpholine? Our diet contains many amines that can be nitrosated. Proline occurs naturally in meat and yields nitro-soproline. Fish contain numerous amines. Vegetables are not benign in this business either. They are high in nitrates, which the body converts to nitrites, which in turn react with amines to form nitrosamines. Basically, the message is that carcinogens are everywhere! We cannot possibly avoid them.

Take coffee as an example. It contains hundreds of compounds, twenty-six of which have been tested for carcinogenicity in rodents. Nineteen of these compounds were positive in at least one test. A typical cup of coffee contains about 10 milligrams of known carcinogens, such as caffeic acid, benzene, styrene, formaldehyde, and furfural. Yet no epidemiological study has shown that coffee increases the risk of cancer in humans. That's because feeding huge amounts of a chemical to a mouse or a rat over the short term may not be a predictor of what happens when humans ingest the same substances in minute doses over the long term in the context of a varied diet.

In order to evaluate human risk, we have to consider not only the carcinogenic potential of a substance but also the degree of exposure. We worry about polycyclic aromatic compounds in burned food and in diesel exhaust because our exposure to these substances over a lifetime can be significant. But we need not be concerned that black pepper contains piperine, which, like morpholine, can be converted to a carcinogen. Its percentage by weight in pepper is trivial, and black pepper is an insignificant portion of our overall diet.

Apples themselves contain furfural and formaldehyde, both of which are recognized carcinogens. But nobody in their right mind would ever suggest that apples cause cancer. That's because these compounds are present in tiny amounts and are greatly outweighed by the flavonoids, vitamins, and fiber found in apples, all of which have established health benefits. Epidemiological studies that attest to the advantages of a fruit and vegetable–oriented diet are published almost weekly. Anyone who reduces their intake of fruit based on an irrational worry about ingesting trace amounts of morpholine is doing themself a great disservice. Personally, I try to eat apples every day. And I'm not concerned about the wax. That's because I tend to believe that there is a difference between scientific toxicology and fairy tales.

The Challenge of Menopause

If I were a woman, I don't know what I would do. About menopause, I mean. You would think that by now scientists would have amassed some pretty conclusive information about how to best deal with a situation that affects half the world's population directly and the other half indirectly. Alas, that is not the case. It seems there is no simple way to deal with declining estrogen levels and the associated mood swings, hot flashes, and reduction in vaginal lubrication. Then there is also the worry that reduced estrogen production equates to an increased risk of heart disease and osteoporosis. We used to think we had a pretty pat answer for these problems. If a lack of hormones cause the problems, why not just replace the hormones? And so millions of women made the attempt to fend off the effects of aging with hormone replacement therapy (HRT).

By and large, women were mostly satisfied with HRT, at least as far as menopausal symptoms went. But in the back of many minds there was a lingering concern about a link between estrogen supplements and breast cancer. After all, there was no doubt that estrogen increased the risk of uterine cancer. This risk, though, could be greatly reduced by taking estrogen together with a progestin, another type of female hormone. But was it possible that estrogen supplements could increase the risk of breast cancer as well? Recent studies suggest that this may indeed be the case, especially with long-term use. To further compound concerns about HRT, not only has the anticipated protection against strokes and heart disease not materialized, but it seems that estrogen supplements may actually have the opposite effect, although to a very small extent.

With anxiety about estrogen increasing, it really is not surprising that many women are looking at "alternative" treatments for menopause. Of course, even the use of the word "treatment" is controversial in this regard because it implies that menopause is a medical condition that requires therapy. More and more women are suggesting the "change of life" is just that: a part of life. They maintain that menopause must be accepted and dealt with by altering lifestyle patterns instead of resorting to drug interventions. They point out that only 10 to 20 percent of Asian women complain of hot flashes compared with 70 to 80 percent of women in Western countries. Some of this discrepancy may be explained by the fact that in many Asian countries, topics such as menopause are just not discussed. However, we also have to consider the possibility that lifestyle patterns play a role.

The most obvious scientific connection to Asian women's lack of hot flashes is soy, a staple in the Asian diet. Isoflavones present in soy are known to have mild estrogenic effects and could, in theory, mitigate the symptoms of menopause. Since the breast cancer rate in Asia is only about one-quarter of North America's, it seems most unlikely that the phytoestrogens present in soy would have the same kind of connection to breast cancer as those found in prescription estrogen supplements. Of course, this does not mean that isoflavones extracted from soy or other sources and put into a pill form are also benign. Recent medical literature is replete with examples of substances isolated from natural sources that have been loudly promoted as dietary supplements and have failed to deliver the goods. For example, the same study that identified the potential harm posed by HRT with regard to cardiovascular disease also showed that vitamin E and vitamin C supplements in post-menopausal women had no benefit in preventing cardiovascular disease and were actually associated with an increased risk—certainly an unexpected finding.

Worries about HRT have oiled the industrial machine that cranks out "natural therapies" for menopause. Many women who leapt off the estrogen bandwagon are now climbing onto the natural therapy one. But just how sturdy a wagon is it? For advice let's turn to the US National Center for Complementary and Alternative Medicine (NCCAM). This organization was created by the Department of Health and Human Services and makes up part of the National Institutes of Health. It was formed in response to the public's demand for more support of alternative treatments and has as its mission the task of fostering rigorous research in such areas. The center is staffed by physicians and scientists who, in general, have a favorable view of unorthodox therapies and can hardly be considered mouthpieces for the pharmaceutical industry. In 2002, NCCAM funded a large-scale survey of the scientific literature on complementary and alternative medicine for menopausal symptoms that encompassed material from 1966 to the present. The results were published in the *Annals of Internal Medicine,* a prestigious peer-reviewed journal.

First let's deal with the stuff the researchers found *didn't* work. Vitamin E at a dose of 400 international units (IU) per day was no better in combating hot flashes than a placebo; it had a 33 percent response, comparable to that seen for many other conditions. Ginseng, used commonly as a tonic against all sorts of ailments in Asia, fared no better. There were even a few reports of postmenopausal bleeding linked to ginseng ingestion. Dong quai, a Chinese herb traditionally prescribed for "female complaints," did not reduce hot flashes. Nor did evening primrose oil. The researchers also looked at studies on creams containing progesterone. Advocates of these products maintain that an increase in blood progesterone can counter symptoms caused by a decline in estrogen. There is some evidence for this claim, but not when creams made from wild yam are involved.

The wild yam argument usually goes like this: Yams contain a compound called "diosgenin" (true), which can be converted to progesterone (also true), and progesterone can help with menopausal symptoms (probably true). What proponents of this theory don't mention is that diosgenin can only be converted to progesterone in the laboratory, not in the body. Not surprisingly, then, no evidence was found that wild yam creams have any greater effect than placebos. Creams that contain actual progesterone synthesized from yams are also on the market. In this case, some studies have shown an improvement in hot flashes when creams containing about 20 milligrams of progesterone were applied daily. But long-term safety studies of these products are not available, and the fact that some women experience postmenopausal bleeding when using topical progesterone is of some concern. Where does all this research leave us? If women are worried about HRT and the alternative treatments mentioned here provide no comfort, what should they do? How much scientific basis is there for the glowing accounts in the popular literature about the benefits of soy, flax, black cohosh, or red clover? Let's investigate.

The dietary intake of soy in China, Japan, and Korea is far higher than in North America, and the frequency of reported menopausal symptoms is far lower. Of course, this observation does not mean soy is responsible for the infrequency of symptoms, but it is an intriguing possibility given that soy certainly contains compounds that can mimic the behavior of estrogens. The compounds of interest are genistein, daidzein, and equol, all members of the isoflavone family. Theses compounds are not found in their free form in soybeans but rather are released by the action of intestinal bacteria, after which they can be absorbed into the bloodstream. So much for the theory. What about the facts?

Studies that investigate soy fall into two basic categories: those that add anywhere from 20 to 60 grams of soy protein per day to the diet and those that use isolated isoflavones in a pill form. These pills contain 30 to 150 milligrams of combined isoflavones, which is roughly equivalent to the amount found in the protein supplements. Results for supplementing the diet with soy protein range from no effect at all to a significant reduction in the frequency of hot flashes. Unfortunately, there are problems with making recommendations based on these studies. Variables such as dosage, product purity, evaluation of hot flashes, and length of study vary widely. For example, the longest study to date, completed in the us, compared 40 grams per day of isoflavone-rich soy protein with the same amount of isoflavone-poor protein as well as with whey protein. There were no significant differences between the groups in the frequency or severity of hot flashes or night sweats. On the other hand, a similar study carried out in Italy using roughly the same amount of soy protein found that hot flashes were reduced by 45 percent while the reduction in the control group, which was given milk protein, was 31 percent. If you want a reduction is hot flashes from eating soy protein, I guess you have to do it in Italy.

The story with isoflavone supplements is similar. Some studies show positive results, but most do not. Even when benefits are detected, they often vanish after six weeks. Furthermore, there is the added concern that no long-term studies on the safety of isoflavone extracts are available. Nor do we have a clear picture of how these substances interact with prescription medications such as tamoxifen, which is used by many breast-cancer patients. The bottom line here is that while incorporating soybeans and bean products into the diet has a long record of safety and may help menopausal symptoms, the safety and efficacy of isoflavone supplements has not been established.

Flaxseeds also contain estrogen precursors. This time the ingredients of interest are lignans, which, again through the action of bacteria in the intestines, release enterolactone and enterodiol, compounds with estrogenic properties. Here, too, there is a lack of consistency in the studies due to the amount and type of flaxseed used and the number of subjects tested. Perhaps the most impressive study was carried out at Laval University in Quebec, Canada. It found that 40 grams of flaxseed daily was as effective at reducing menopausal symptoms as HRT. That's a lot of flaxseed—about 5 tablespoons' worth. Let's remember that lignans are effective against hot flashes because they are estrogenic. Could they not cause a problem if flaxseeds are consumed over the long term? Nobody knows. And what about women who are taking tamoxifen? They're probably better off playing it safe and limiting flax intake.

Now we come to black cohosh, a herbal product commonly available as an ingredient in Remifemin, a supplement for menopause symptoms. As is the case with any natural product, black cohosh contains dozens of compounds with potential for biological activity. The latest research suggests that these compounds have only weak estrogenic potential, so whatever benefit black cohosh provides comes via some other unidentified mechanism. Again, some studies show that women taking black cohosh experience improvement of menopausal symptoms over the short term. As the longest clinical trials have lasted only six months, the long-term effects of black cohosh are unknown. No human trials have been published about the effects of black cohosh on breast or uterine tissue, and there is no evidence that it can protect against osteoporosis, which is one of the benefits of HRT. It must be remembered that *herbal* does not equate to *safe*. Recently, researchers at Duquesne University in Pennsylvania fed black cohosh to female mice that were genetically prone to breast cancer. While there was no increase in the incidence of breast cancer, if cancer did develop, it spread much more quickly in the animals treated with black cohosh. Does this result mean that a woman with undiagnosed breast cancer may be at greater risk if she takes black cohosh? Nobody knows.

Nobody knows if women stand to benefit from red clover extract either. Like soy, red clover contains phytoestrogens, including biochanin A and formononetin, which are not found in soy. Many women claim that their menopausal symptoms are eased when they take Promensil, a standardized red clover preparation. The latest scientific study on this product, however, does not support this view. When 250 postmenopausal women were treated with Promensil or placebo for twelve weeks, researchers found no clinically important effect on hot flashes or other menopausal symptoms. Interestingly, the study was funded by Novogen, the manufacturer of Promensil. On the other hand, a British study of 29 postmenopausal women over eight weeks found that vaginal dryness, a cause of painful intercourse, was alleviated with the use of the same product.

Confusing, isn't it? That's why I said at the outset that I don't know what I would do if I had to deal with menopause. For severe symptoms, hormone replacement therapy still has its place. I think, though, that I would try eating more soy foods, give black cohosh a six-month trial, experiment with Promensil if need be, and exercise like mad.

CARPETS AND CHEMICAL SENSITIVITIES

The book has all the elements of a great movie. There are incompetent physicians, crooked judges, unethical lawyers, greedy companies, corrupted experts, bizarre conspiracies, innocent victims, and terrorists with chemical weapons. But these terrorists don't hijack airplanes or carry out suicide bombings. Instead, they find ways to infiltrate our schools, homes, and offices, where they can release their toxins and wreak havoc with our health. We have a common name for these terrorists. We call them "carpets."

Toxic Carpet is a remarkable book. In it, Glenn Beebe recounts how his family's health was shredded by an office carpet and describes his futile battles for compensation. Back in 1980 the Beebes ran a small family business in Cincinnati, Ohio, and life was flowing along smoothly until a new carpet was installed in the office. As is usually the case, the new carpet had a smell, but to the Beebes it seemed more pungent than normal. The odor persisted, and soon they began to experience headaches, dry mouth, burning eyes, shortness of breath, fatigue, and tingling sensations. They became really concerned when dead spiders, still in a standing position, began to appear on the carpet. Neither the store where the carpet was purchased nor the manufacturer offered any sort of explanation or compensation. While the manufacturer did admit that chemicals "outgas" from new carpeting, it claimed tests had shown that in the concentrations present they were "safe." If there was an air-quality problem, the carpetmaker said, it was probably due to chemicals emanating from the fiberglass ductwork in the office.

Nonsense, the Beebes replied. After they removed the offensive carpet and replaced it with another, the problems in the office subsided. However, whenever they were exposed to smells, their symptoms returned with a vengeance. Diesel fumes, perfumes, paints, solvents, and even newsprint triggered respiratory problems, chest pains, and memory lapses. They were experiencing "multiple chemical sensitivity." The carpet had destroyed their lives, so the Beebes decided to sue. As described in *Toxic Carpet,* that process turned out to be quite an adventure. Physicians labeled the Beebes hypochondriacs, blood tests disappeared, witnesses vanished, lawyers milked them dry, and even the judge was against them. The courtroom in which the trial was held had just been remodeled and was permeated with all sorts of smells that interfered with the Beebes' concentration. The judge, claiming he couldn't hear the proceedings properly, ordered an air conditioner to be turned off, forcing the Beebes to wear respirators, which impeded the delivery of their arguments. Before the case went to the jury, the manufacturer offered a settlement of $100,000, which the Beebes refused. The jury took just forty-five minutes to return a verdict in favor of the defendants. Whatever problem the Beebes had, the message seemed to be that it wasn't caused by broadloom.

The plaintiffs were outraged by the decision, and Mr. Beebe offered the opinion that letting the manufacturer off "scot-free" was no different than turning a criminal loose to repeat the same crime. But he decided there was no point in appealing because "money and power control our court system." Instead, Glenn Beebe decided to write a book that exposed the dangers of toxic carpets and alerted the public to the existence of multiple chemical sensitivity.

I'm not sure what the truth is regarding the Beebes' allegations. But I can tell you that multiple chemical sensitivity is a real ailment. The condition may have a psychological component, but that does not make it any less real. For whatever reason, the victims of the condition do suffer. The scenario often unfolds as it did for the Beebes: an initial exposure to a noticeable smell triggers a cascade of events that sometimes culminates in the victim being forced to lead a life of isolation from the odors found everywhere in our modern lives. Such an extreme chemical sensitivity is rare, and it would be unusual, though certainly not impossible, for a husband and wife to be simultaneously affected.

There is, of course, no question that new carpets smell. The odor is mostly due to 4-phenylcyclohexene, an undesired by-product in the manufacture of styrene butadiene rubber, which is used in the backing of most carpets. Numerous other compounds used in the manufacture of synthetic carpet also outgas, but with ventilation the smell usually dissipates within a short time. Without a doubt some people experience eye, nose, and respiratory tract irritation from new carpet smell, but that doesn't guarantee that their lives are about to crumble as the Beebes claim theirs did. Actually, great improvements have been made by manufacturers to cut down on outgassing, and a number of "green" carpets that give off practically no smell are available. You can even buy carpets woven from the wool of organically raised sheep—whatever that means.

Frankly, aside from isolated cases, I don't think new carpet smell is a plague. But that doesn't mean the issue is of no concern. Carpets do serve as a repository for all sorts of chemicals. Benzopyrene, a nasty carcinogen found in tobacco smoke and cooking fumes, settles on carpets along with residues from air fresheners, cleaning agents, and pesticides. The latter are usually tracked into a home from outside. Humans who wear shoes in the house, and animals that do not, increase the pesticide load in carpet dust some 400-fold! So get a good doormat and make everyone take off their shoes before walking on the carpet! Then there is the matter of dust mites and their poop. These tiny spider-like creatures like to dine and defecate in carpets. Their feces, which can be highly allergenic, are part of carpet dust. Unfortunately, this dust doesn't stay put. It has a "grasshopper effect": disturbing the carpet stirs up the dust and leads to the contamination of toys and even food.

What's the answer? Vacuum and then vacuum some more! Use a machine that has been shown to minimize the emission of small particles in the exhaust. And you are not going to like the next bit of advice. Studies have shown that every week, twenty-five passes of a vacuum are needed to clean a rug within 4 feet of a home's main entrance, and sixteen passes are needed for heavily trafficked areas. I could tell you more interesting facts about carpets, but I'd better start vacuuming . . .

Adventures in the Health Food Store

"My mother hasn't been dieting but has lost about twenty pounds in a month and she feels tired all the time. What can you recommend?" That's the question my students were instructed to ask when approached by a salesperson in a health food store. It was part of an assignment designed to investigate the reliability of advice provided by such establishments. Of course, there is only one reasonable answer to such a question, which is: "Tell your mother she should see a physician." And that was the answer given in about 50 percent of the cases. But the rest of the salespeople offered up a bewildering array of vitamins, "glandulars," protein powders, creatine supplements, chromium pills, exotic juices, magnetic bracelets, and drops of "aerobic oxygen." One salesperson advised that the mother immediately start drinking only distilled water and eating only organic produce. Another came up with a truly startling diagnosis for the weight loss: "I bet your mother has a microwave oven in her kitchen and uses Teflon pans."

Although this adventure was to be only a learning experience for students, the results were so interesting that I thought they merited publication. Unfortunately, since we had not planned on a rigorous study, the students had not documented the "evidence" to a degree that would pass scientific muster and we had not considered what an ethics committee might say about our project. So you can imagine my interest when I saw a recent paper in the peer-reviewed publication *Breast Cancer Research* with the title "Health food store recommendations: Implications for breast cancer patients." Edward Mills and his colleagues had sent research associates of varying ages into health food stores in the Toronto, Canada area to browse the shelves until approached by an employee. They would then reveal that their mother had been diagnosed with breast cancer and proceed to ask for suggestions about what to do. The study had been approved by Mills's ethics committee and his research associates had been schooled in what questions to ask and what information to divulge. Each encounter was documented immediately after the researcher left the store.

Altogether, thirty-four stores were visited. In twenty-seven of these visits, recommendations were made for the use of some sort of "natural health product." A total of thirty-three different treatments were suggested, including the usual vitamins, a host of botanicals, shark cartilage, garlic, grape seed extract, dehydrated vegetables, antioxidants, herbal teas, and, in one case, a preparation made from the Venus fly trap. Only one common feature links all these products: a lack of scientific efficacy for the treatment of breast cancer. The most commonly recommended product was Essiac, an herbal remedy that contains burdock, Indian rhubarb, sorrel, and slippery elm. It was first popularized in the 1930s by Rene Caisse (*Essiac* is her name spelled backward), a Canadian nurse who claimed to have learned of this cancer cure from Ojibway medicine men. Essiac has actually been tested both in humans and animals and has shown no antitumor activity.

Less than one-quarter of the employees in the study discussed the potential for any interaction between prescription drugs and natural health products, and only three employees mentioned the possibility of any adverse effects. Two employees claimed that their recommended products could cure breast cancer, and perhaps most disturbing was the suggestion by one salesperson that tamoxifen, an established breast cancer treatment, be discontinued. The recommendations did not come cheap; on average, the products suggested would set the patient back $58 a month, with some costing as much as $600. Finally, only 44 percent of the employees encountered recommended visiting a health care professional, and even then the majority suggested a visit to a naturopath.

By now some of you may be thinking that I'm reporting on a study that was carried out by some medical or pharmaceutical establishment–types bent on showing the frailties of the natural health movement. As the argument often goes, these people are worried that their "slash, burn, and poison" methods used to treat cancer are going to be replaced by gentle, effective natural approaches. Profits are at stake, so they supposedly design studies that will discredit the "opposition." Well, this particular study was neither funded nor carried out by physicians or drug companies. Edward Mills, the lead author, is the director of research at the Canadian College of Naturopathic Medicine. He, like many others involved in the pursuit of effective natural products, is disturbed by the unreliable and sometimes dangerous advice given in health food stores. Spreading nonsense certainly does not further the cause of complementary and alternative medicines.

The ideal way to cut down on the confusion that surrounds the treatment of breast cancer is to reduce the need for any treatment. In other words, prevent the disease from occurring in the first place. Presently, there is no consensus on the relationship between diet and breast cancer, but being overweight is an established risk factor. This is probably the case because fatty tissues are not only capable of storing estrogen but are also adept at converting male hormones produced by the ovaries and adrenal glands into estrogen. High levels of this female hormone have been conclusively linked to some breast cancers.

Estrogen production can also be reduced with physical activity. Postmenopausal women who exercise moderately for roughly two hours per week can reduce their risk of breast cancer by as much as 20 percent. Younger women who work out for at least four hours per week during their reproductive years can reduce it by 50 percent! Exercise works for young girls as well. We know that early onset of menstruation is linked to more intense hormone exposure throughout life and therefore to breast cancer. A large study of elementary school girls found that just five hours of exercise per week can delay puberty and thereby lower the threat of the disease. Exercise is good for you. And that advice is based on science, which makes it a lot more reliable than what you may hear in a health food store. Just ask the director of research at the Canadian College of Naturopathic Medicine.

The Mystery of Alzheimer's Disease

What an interesting patient! "Auguste D" was only fifty-one years old when she was admitted to a hospital in Frankfurt, Germany, in 1901, seemingly suffering from senile dementia. She was confused, her memory was impaired, and her behavior was unpredictable. When asked what she was eating after being served pork and cauliflower for lunch, she replied, "Spinach." The patient then went on to offer a surprisingly insightful self-diagnosis: "I have lost myself," she explained. It's little wonder that this unusual case captivated the imagination of attending physician Dr. Alois Alzheimer. He followed Auguste's deterioration carefully for five years, wondering what was going on in her eroding brain.

When Auguste died in 1906, Dr. Alzheimer got his chance to find out. His postmortem microscopic examination revealed that her brain had undergone some very peculiar changes. A sort of sticky material had deposited in the spaces between the nerve cells, some of which harbored tiny twisted fibers. Just a few months after the patient's death, at a conference of German psychiatrists, Alzheimer described the "neurofibrillary tangles" and "senile plaques" that have since become the hallmark of Alzheimer's disease.

Simply put, with Alzheimer's disease, the brain's machinery becomes "gummed up." In a healthy brain, nerve cells communicate with each other through the release of special chemicals called neurotransmitters, but this communication is impaired by the presence of tangles and plaques. As these abnormal structures accumulate, nerve cells die, the brain shrinks, and mental and physical incapacity sets in. Once autopsies of several of Alzheimer's patients had confirmed the presence of plaques and tangles as a characteristic feature of the disease, obvious questions arose: What are these abnormal deposits and why do they form? The first question has now been answered. Both the sticky plaques and tangles result from accumulations of proteins—beta-amyloid in the case of plaques, and an abnormal, insoluble form of a protein called "tau" in the case of the neurofibrillary tangles. The "why" question has been far more difficult to answer. Some genes predispose people to Alzheimer's, but most patients suffering from this disease do not have these genes. Only age has been shown to be a clear-cut risk factor in the disease, which afflicts about 1 in 10 people over the age of sixty-five and 1 in 2 over eighty-five. To complicate matters further, not all people with plaques and tangles in their brains show symptoms of Alzheimer's disease. We are certain of this fact thanks to the remarkable Nun Study initiated in 1991 by David Snowdon of the University of Kentucky.

Close to 700 Catholic nuns in several American convents, most of whom had careers as teachers, were asked if they wanted to keep teaching even after they died. The sisters welcomed the opportunity to further the cause of science and agreed to donate their brains to help researchers learn about the aging process. Much has already been learned. One nun who died of a heart attack at age eighty-five had not only a genetic marker for Alzheimer's but also such extensive plaques and tangles in her brain that she should have been totally incapacitated. Yet her cognitive abilities showed no deterioration at all. Somehow, her brain was resistant to the ravages of the disease! Eventually the researchers found that about one-third of the nuns with severe deposits had escaped mental deterioration. What had protected them? It seemed it wasn't anything they had; it was what they *didn't* have. Their brains showed no history of the small strokes common to the elderly. By contrast, the nuns who had plaques and tangles *and* Alzheimer's symptoms also had a history of ministrokes. Apparently abnormal protein deposits are required but alone are not sufficient for a manifestation of Alzheimer's disease. This is an interesting finding because it suggests that control of blood pressure and cholesterol levels may reduce the risk of Alzheimer's. Some studies have already shown that patients taking "statin drugs" to lower cholesterol are less prone to the disease.

What else have we learned about the disease? Senior women who are overweight (but surprisingly, not senior men) are more likely to develop Alzheimer's. Women over the age of sixty-five who take combined estrogen and progestin are at greater risk. But if they are also prescribed testosterone supplements, they are at reduced risk. People who take anti-inflammatory drugs for long periods appear to be less threatened by Alzheimer's, probably because inflammation in the brain exacerbates the effects of the plaques and tangles. Interestingly, the disease is less prevalent in India, and some researchers have suggested that the heavy consumption of curcumin, an ingredient in yellow curry, is responsible. This extract of the root of the *Curcuma longa* plant has been shown to have anti-inflammatory effects in laboratory studies.

Other aspects of diet have also been examined. Perhaps the most interesting finding has been the association between high levels of homocysteine in the blood and Alzheimer's disease. Homocysteine forms from methionine, an essential amino acid in the diet, and builds up in the bloodstream if the B vitamins required to metabolize it are not present in adequate amounts. Indeed, the Nun Study has shown quite a strong link between low levels of folic acid, one of the B vitamins, and increased risk of Alzheimer's. A diet without an excess of methionine and one high in the B vitamins should therefore offer some protection. Meat is very rich in methionine, and whole grains and leafy greens are an excellent source of folic acid. Maybe Auguste D, the original Alzheimer's patient who said she was eating spinach when she was really eating pork, had some sort of message for us. And I think we can also learn something from Dr. Alzheimer himself. He used to walk around the lab where his students pored through microscopes with a cigar constantly dangling from his mouth. By the end of the day, every student's workplace was adorned with a standing cigar stump. Well, it turns out that smoking roughly doubles the risk of Alzheimer's disease, probably by generating free radicals that attack brain cells. But Alzheimer did not succumb to the disease that bears his name; he died of heart disease, another smoking-related ailment. At the time he was fifty-one, coincidentally the same age as his famous patient, Auguste D, when she was admitted to the hospital.

When it comes to thinking about treating Alzheimer's disease, we have to realize that the most complex computer in the world is a simple machine when compared with the human brain. How is it that Mozart composed a symphony at age five when I can hardly tell one note from another? How is it that some people can have IQs so low that they are unable to fend for themselves but can nevertheless perform astounding mathematical feats (like Dustin Hoffman's character in the movie *Rain Man*, for example)? How is it that Audrey Cyr, an eighty-six-year-old grandmother with no previous artistic flair, lost her mental capabilities but blossomed into a renowned painter after she was afflicted with Alzheimer's disease? Modern science has no answers to these questions. Still, we aren't completely at a loss when it comes to dealing with mental problems. The anatomy of the brain has been studied in great detail and neuroscientists can tell us which areas control language skills, spatial perception, or short-term memory. But what exactly goes on in those areas is more of a mystery. Certainly, chemistry is somehow involved. We know, for example, that levels of serotonin or phenylethylamine in the brain can affect mood, that a lack of dopamine can result in Parkinson's disease, and that depleted levels of acetylcholine are associated with Alzheimer's disease. We also know that medications that raise serotonin levels (e.g., serotonin reuptake inhibitors such as Prozac) or dopamine concentrations (e.g., Levodopa) have been useful in the treatment of depression and Parkinson's disease. It seems reasonable, then, to theorize that drugs capable of enhancing acetylcholine activity should prove useful in the management of Alzheimer's disease.

Acetylcholine is one of many neurotransmitters that nerve cells use to communicate with each other. These cells are not in actual physical contact. Rather, when one is stimulated, it releases a neurotransmitter into the tiny gap, or synapse, that separates the nerve cells. The neurotransmitter then binds to receptors in an adjacent nerve cell, causing stimulation. After it has delivered its message, the neurotransmitter is broken down by specific enzymes to prevent overstimulation. As an example, just think of what happens when you put your hand on a hot stove. You don't have to think about removing it—you do it reflexively. That's because a neurotransmitter-mediated message has gone to your brain, which has quickly responded in like fashion. Of course, you don't keep jerking your hand once you have removed it from the burner. The neurotransmitters have done their job and have been broken down. It should come as no surprise, then, that interference with such activity can result in impaired brain function. And it seems that the buildup of neurofibrillary tangles and senile plaques in and around nerve cells, as happens in Alzheimer's disease, results in precisely this kind of interference. Why not try to raise acetylcholine levels in the brains of Alzheimer's patients? Actually, this is exactly what physicians do when they prescribe cholinesterase inhibitors, the main drugs used in the management of Alzheimer's disease.

Tacrine (Cognex), donepezil (Aricept), rivastigmine (Exelon), and galantamine (Reminyl) all inactivate to some degree the enzyme that metabolizes acetylcholine. These drugs do not perform miracles, but patients with mild to moderate Alzheimer's disease can experience temporary improvement in memory and in the capacity to carry out routine activities such as dressing while taking them. Typically, the progression of symptoms is delayed by several months. An added benefit of the medication is a reduction in the aggressiveness and agitation that often accompanies Alzheimer's. If these behavioral problems persist, or if patients experience delusions and hallucinations, which unfortunately are common in Alzheimer's, antipsychotic medications such as haloperidol (Haldol) or risperidone (Risperdal) are sometimes employed. Since depression is often coupled with Alzheimer's disease, antidepressants such as paroxetine (Paxil) or sertraline (Zoloft) can make it into the mix. Memantine, a medication that has been used in Europe for a decade, may become the first drug in North America to be approved for the treatment of severe Alzheimer's dementia. Memantine has a novel mechanism of action: it blocks the action of glutamate, another neurotransmitter that in excessive amounts is known to damage nerve cells. Again, the hopes are modest. Studies show that memantine can delay the progress of the disease by several months.

Anytime one interferes with activity in the nervous system, side effects are to be expected. Such is the case with all the medications mentioned above, with problems ranging from nausea and vomiting to uncontrolled movements. Kinder, gentler, and more effective treatments are obviously desired but have proven elusive. Large doses of vitamin E (2,000 IU per day) seemed a reasonable option given the plaques and tangles in the brains of Alzheimer's patients apparently make nerve cells more susceptible to attack by the free radicals generated as a normal by-product of metabolism. Vitamin E is an established antioxidant, or free-radical scavenger. Results have been less than spectacular, but the vitamin has helped to delay the institutionalization of some Alzheimer's patients. A recent and intriguing study suggests that people who take a daily dose of vitamin E (400 IU) and vitamin C (500 milligrams) may be less prone to developing Alzheimer's. Supplements of ginkgo biloba, a substance that increases circulation in the brain, have been tried as well as huperzine, an ancient Chinese remedy for mental problems. Results for both substances were hardly better than results with placebos, and there is the added concern that due to the lack of regulations about "natural health products," we can never be sure of their actual contents.

There have been two recent developments in the Alzheimer's battle. A small study of some 100 patients showed that the antibiotics doxycycline and rifampin were effective in reducing deterioration in mental function over the short term, perhaps by interfering with the accumulation of the amyloid proteins that make up the senile plaques. And researchers in England have followed up on the ideas of the sixteenth-century herbalist John Gerard, who wrote that sage "is singularly good for the head and brain and quickeneth the nerves and memory." They found that in forty-four healthy volunteers, those who took sage-oil capsules performed better on memory tests than those who took placebos. When lead researcher Nicola Tildesley was asked if the results convinced her to take sage oil, she replied that indeed she has begun to do so. Then she quickly added, "Whenever I remember." And so it goes with Alzheimer's research.

Going through a Phase

Stealing silver dishes from the palace kitchen during the reign of King Charles IX of France was not a good idea. Especially if you got caught. As one of the king's sticky-fingered cooks discovered, the penalty was death by hanging. But just as the thief was preparing to meet his maker, he was offered a chance to cheat the gallows. He simply had to submit himself to a little experiment.

Kings and queens in those days lived in constant fear of being poisoned. Indeed, dispensable servants were often asked to taste food before it reached royal lips. So it is not surprising that Charles showed a keen interest when he was approached by a Spanish nobleman claiming to have an antidote against all

poisons. The secret, the nobleman explained, lay in the magical properties of the "bezoar stone."

Bezoars are not mythical; they really do exist. But they are not stones. A bezoar is a calcified mass that can form in the digestive tract of animals when digestive juices harden around some foreign material. Rare substances often spawn myths, and such was the case with bezoars in the 1500s. As the Spaniard told the king, powdered bezoar, especially the variety that came from the intestines of the Persian wild goat, was an effective antidote to all poisons. The king wanted a "scientific" opinion on the matter and sent for his chief military surgeon, Ambroise Paré. The doctor was highly skeptical of a universal antidote, as he believed, quite correctly, that different poisons act in different ways. But when the Spaniard persisted in his claims, Paré proposed an experiment. That's when the king sent for the condemned cook. If the cook agreed to swallow some powdered bezoar followed by a poison, and survived, he would be freed. Of course, the ill-fated cook readily agreed. The royal apothecary was summoned and asked to recommend a particularly potent poison. His suggestion was bichloride of mercury. It turned out to be no match for the mythical powers of the bezoar, and after several hours of terrible agony the cook expired. King Charles ordered the bezoar to be thrown into the fire, but historical accounts are silent about the fate of the Spanish nobleman. One would think that he too had a few agonizing hours—at the very least.

Why discuss the saga of the thief and the bezoar? Because it represents a clear, albeit primitive, example of a clinical trial. A hypothesis was put forward, an experiment was designed to test it, and the results were recorded. Mercifully, today's clinical trials are more sophisticated than this early experiment, but they are still not without their pitfalls. Let's take a look.

The appearance of a new drug in the marketplace is preceded by many years of carefully planned studies. Before any human trials are even contemplated, a new drug is administered to various species of animals to test for potential effectiveness and toxicity. Testing substances on animals raises ethical considerations, but unfortunately there is no alternative—at least not as long as we value human life above animal life. While it is certainly true that a rat is not simply a miniature person, we must have some indication about toxicity before human trials can be considered.

If laboratory experiments and animal studies give reason to believe that a drug may be effective in the treatment of some disease and is unlikely to cause severe side effects, a Phase I trial is initiated. Roughly a few dozen healthy volunteers are recruited and given the drug in various doses. The trial is aimed strictly at determining safety and usually lasts about a year. How the drug is metabolized and excreted is studied, and any side effects that arise are carefully recorded. If the safety profile turns out to be acceptable (as happens with about three-quarters of the drugs that make it to Phase I trials), then it is time to test for efficacy in a Phase II study. This time, several hundred patients with the medical condition the drug is designed to treat are recruited and are divided into an "experimental" group and a "control" group. The experimental group gets the drug while the control group receives a placebo. The ideal study is conducted in a "double blind" fashion, meaning that neither the subjects nor the experimenters are aware of which subject are actually receiving the drug until the study is terminated. Phase II studies can last several years, during which time changes in health status and development of side effects are carefully monitored.

If results from Phase II show significant efficacy and a good safety profile, a Phase III study is designed. For this phase, thousands of patients are recruited and separated into different "cohorts," or groups that have a statistical factor such as age in common. Various doses are studied and efficacy is compared with other drugs already on the market. Potential problems that may arise if the drug is used in combination with other medications are also examined. A Phase III study can last up to five years. If after all this testing the benefits of a new drug seem to clearly outweigh the risks, then all the supporting documentation is submitted to the appropriate government authority (Health Canada, for example, or the us Food and Drug Administration) for approval. By this time, however, researchers have usually already participated in numerous discussions with government scientists, keeping them abreast of the ongoing studies and ensuring that requirements for eventual approval are being met.

Studies of a new drug's efficacy and safety profile do not end with approval to market the drug. When approval is granted, Phase IV studies begin. In this phase, the use of the drug in the general population is monitored by the manufacturer, who is required to report any problems that arise for up to ten years. In rare cases where problems are discovered, the drug is removed from the market. The system is not perfect; none ever will be. There are always concerns about proper monitoring of studies (100,000 trials are conducted annually in North America), risks to subjects, financial conflicts of interest, and the existence of studies that don't really advance the frontiers of medicine. But without a rigid regulatory framework, unproven remedies with their promises to cure all diseases would rule supreme. And without proper studies, we would never have discovered the real antidote to bichloride of mercury. So if anyone ever makes you an offer like the one Charles IX made to his cook, don't ask for a bezoar. Ask for dimercaprol, the scientifically proven antidote.

Russian Spies and Sick Turkeys

So tell me: What is the relationship between moldy corn and Otto Preminger's last movie? The film we're talking about is *The Human Factor,* an adaptation of Graham Greene's 1978 spy novel. In a nutshell, the story is about a suspected mole in the British Foreign Office, who is thought to be leaking information to the Russians. A decision is taken to eliminate the traitor by poisoning him so as not to attract any adverse publicity. Of course, a poison that works swiftly would raise suspicion, so a physician working for the foreign office decides on aflatoxin B, a chemical that triggers cancer of the liver. Pretty ingenious and feasible too—at least in theory.

Aflatoxins are naturally occurring toxins produced by molds of the genus *Aspergillus flavus* that can infect nuts and grains, particularly corn, when they are stored under warm, humid conditions. The aflatoxin story is an interesting one because these compounds were the first recognized dietary carcinogens.

They were not the first carcinogens to be identified; that honor goes to substances found in snuff. In 1761, John Hill, an English physician, noted that snuff users were more likely to develop nasal tumors. In 1775, Percival Potts, a surgeon, discovered that chimney sweeps had an unusually high incidence of skin cancer on the scrotum. Amazingly, chimney sweeps on the continent did not experience this rare cancer. What was the difference? Bathing was more common on the continent than in Britain, and British chimney sweeps walked around continually covered in soot from head to toe. Scientists wondered if there was something in soot that was responsible for the disease. The sticky material inside chimneys, known as creosote, was the likely candidate. Researchers began to explore this possibility by painting creosote on the skin of animals to see if tumors could be induced. For about 150 years these experiments went on sporadically and unsuccessfully.

Then along came Katsusaburo Yamagiwa, a medical researcher in Japan. He had something that his predecessors in this research area apparently did not have: patience. Yamagiwa carefully painted the ears of 137 rabbits with creosote every day for over a year. Finally, tumors began to appear in seven of the rabbits. This result could not have happened by chance; Yamagiwa had found a carcinogen. Actually, he had found many carcinogens. Creosote is a very complex mixture of chemicals, and to this day it has not been completely analyzed. But some of its components, such as the notorious polycyclic aromatic hydrocarbons (PAHs), have been identified as definite carcinogens.

By the time Yamagiwa made his discovery, epidemiologists had realized that some occupations other than chimney sweeping were also associated with higher rates of certain cancers. Dyers had unusually high rates of bladder cancer, and workers using chromium were more likely to develop lung cancer. By the 1930s, the existence of a link between certain industrial

chemicals and cancer had become clear. Obviously, though, not everyone affected by cancer had exposure to these substances, so there had to be other causative agents as well. Genetic factors were clearly important because some cancers, such as those of the breast, were more likely to occur in people who had relatives with the disease. Viruses were also implicated because some of these had been shown to cause cancer in animals. Nobody, though, thought that food could be a factor. At least not until 1960, when turkeys began to die en masse on poultry farms in England.

Within a few months, over 100,000 birds perished from what came to be called "Turkey X disease." When ducks and pheasants began to die as well, scientists looked for a common cause and found it in the feed that the birds were given. The turkeys, ducks, and pheasants all had been fed Brazilian peanut meal. It didn't take long to discover that this food could induce Turkey X disease in healthy animals. Within a year, the culprit was identified as the mold *Aspergillus flavus,* which produced the appropriately named "aflatoxins."

The turkey episode raised questions about a potential risk to humans who consumed aflatoxin-contaminated food. The concern became very real when it was discovered that crops damaged by drought, which was not an unusual occurrence, were particularly prone to contamination. And the red flag was raised high when laboratory studies showed that aflatoxins could induce cancer. It didn't take long for scientists to discover that this relationship existed outside the laboratory as well. Epidemiological surveys confirmed a high rate of liver cancer in areas of high aflatoxin exposure, particularly China, where rice often is contaminated by molds. As many as 1 in 10 Chinese adults develop liver cancer. Studies in Africa confirmed the theory. In both Kenya and Mozambique peanut meal is a staple of the diet, but storage conditions in the two countries differ. Calculations show that in Kenya the average daily intake of aflatoxins is 3.5 nanograms per kilograms of body weight (a nanogram is one-billionth of a gram) while in Mozambique it is 220 nanograms per kilogram. The incidence of liver cancer in Mozambique is thirteen times higher than in Kenya. Further studies revealed that exposure to aflatoxins was even more worrisome in people with hepatitis B, whose livers were more likely to be affected by aflatoxins.

In North America, stored grains are routinely assayed for aflatoxins. For corn, a common method of inspection is shining greenish yellow light on the kernels, which fluoresce if they are contaminated with *Aspergillus flavus.* A fascinating new technique for detecting the mold uses sound. Corn kernels are heated by infrared radiation and are then quickly cooled. Those that are contaminated make a different sound when they contract, and computer analysis of the frequencies produced can determine the extent of contamination. If *Aspergillus* is found a laboratory analysis is carried out, and corn with aflatoxin levels greater than 20 parts per billion is deemed unfit for consumption. Of course, these tests do not mean that we have no exposure to aflatoxin because obviously not every lot of every grain can be tested. Nobody knows how many cases of liver cancer in North America might be attributed to aflatoxins. Judging by a fascinating case of attempted suicide, probably not many. A young American woman tried to kill herself by ingesting 5.5 milligrams of aflatoxin B (a high dose) that she had stolen from a laboratory. Nothing happened. Six months later, she tried again with a total of 35 milligrams over two weeks. Again, nothing. She then decided life was worth living after all, and fourteen years later she was still in good health, with no liver problems. Oh well, it worked in *The Human Factor*.

The Strange Case of the Stinky Man

What an awful, putrid odor it was! The physician had never smelled anything like it. In the small examination room, the stench coming from the man's arm was virtually intolerable. Amazingly, the arm looked normal except for an inflamed finger. That's where the man, who worked in a chicken processing plant, had pricked himself with a chicken bone. Antibiotics were administered immediately, but to no avail. The redness in the finger persisted and the man's odor got worse and worse. Skin biopsy samples from the arm revealed the presence of *Clostridium novyi* bacteria. Despite the sensitivity of the bacteria to antibiotics in the laboratory, no antibiotic was able to eradicate it from the unfortunate man's arm. Desperate physicians tried everything. They exposed the unfortunate victim to ultraviolet light. They put him in a hyperbaric oxygen chamber. They wrapped him in giant odor eaters. Nothing worked. The poor soul eventually had to give up his job and social life because people literally recoiled with horror when he walked into a room.

What was going on here? To get a hint, doctors had to isolate the compounds responsible for the smell. They captured the vapors being released by the man and subjected them to analysis by gas-liquid chromatography, a technique that can separate the components of a mixture. After separation, the individual components of the odor were determined to be acetic, propionic, butyric, and 4-methylvaleric acids. Acetic acid is not particularly disturbing; after all, this is the substance found in vinegar. Propionic and butyric acids are more revolting, as they are found in the fragrance of rancid butter. But as far as horrific smells go, valeric acid takes the cake. It is reminiscent of the smell of manure. If you want to get just an idea of what it smells like, sniff some Limburger cheese and imagine that odor magnified thousands of times. The flavor and scent of Limburger are the result of compounds produced by specific bacteria added to the cheese. One of the products of this bacterial metabolism is valeric acid. The *Clostridium novyi* bacteria also produces this fragrance, as our unlucky chicken processor discovered. Poultry can be contaminated with this bacteria, but transfer to humans is extremely rare. Nobody knows why it happened in this case and why antibiotics were unable to wipe the bacteria out. The victim did not have a depressed immune system and had no other medical conditions. For six years he suffered, living a life not worth living. Then, all of a sudden, the bacteria and the odor it produced just disappeared! What a mysterious machine the body is!

Although the smelly chicken plucker appears to be a unique case in the annals of science, Greek mythology may provide a precedent. Philoctetes was one of the Argonauts assembled by Jason to help in the search for the Golden Fleece. On the way to Troy, the Greek fleet called at the island of Lemnos, where Philoctetes was bitten on the foot by a poisonous snake. The wound gave off a terrible smell and Philoctetes was abandoned by his fellow Greeks. Mythology, as the name implies, is based on myths, but myths are often prompted by real-life experiences. Maybe the mythmaker in this case had noted some horrific smell coming from a snake bite. And who knows—perhaps some snakes, like chickens, harbor *Clostridium novyi.*

Chickens are not the only potential cause of body odor problems. Fish can do it too. And fish as the root of the problem is a far more common situation. As many as 1 in 10,000 people may be affected to some extent by trimethylaminuria, or, as it is better known, "fish odor syndrome." Anyone who has ever worked in an organic chemistry lab will recall, without any fondness, the smell of trimethylamine. But you needn't have worked in a lab to have experienced the smell, which is the putrid odor of rotting fish. People affected by fish odor syndrome excrete the compound in their breath, sweat, saliva, vaginal secretions, and urine. They inevitably suffer social isolation and often become seriously depressed. And it's all the fault of a bit of biochemistry gone awry.

The culprits in this condition are two dietary components, namely choline and carnitine. Choline is found in eggs, legumes, nuts, meats, some vegetables, as well as fish. It is an essential nutrient required for the synthesis of cell membranes, and it is also the body's precursor for acetylcholine, an important neurotransmitter. Carnitine is found mostly in meats and is required by cells to help in the burning of fats for energy. Much of the choline and carnitine found in food is absorbed into the bloodstream and is delivered to cells as required. But some is broken down by bacteria in the intestine in a process that releases trimethylamine. For the majority of the population, this process is of no concern because the trimethylamine is converted in the liver to odorless trimethylamine oxide, which is then excreted in the urine. But for those with fish odor syndrome, not enough of the enzyme (trimethylamine oxidase) needed to carry out this conversion in the liver is produced, thanks to a faulty gene inherited from both parents known as a "recessive trait." The result is a buildup of trimethylamine, which then escapes from the body through various secretions, wreaking havoc with victims' lives as well as that the lives of those around them. Avoiding foods rich in choline and carnitine is the only way to curb the odor.

Fish odor syndrome has a broad spectrum of effects. Curiously, sometimes only the patient seems to be aware of the problem. Witness the case of a teenager who began to notice a strange and unpleasant smell, "something like rotten fish," in her mouth when she woke up in the morning. As she got older the problem became worse and worse. She consulted doctors, none of whom could smell anything. CAT scans and MRI examinations revealed nothing. Even her tonsils were removed on the hunch that the problem was caused by food particles sticking to them. The poor young woman became more and more depressed. Finally, a clue surfaced when she mentioned that she noticed the smell only when she exhaled. She was given guaifenesin, an expectorant to increase secretions, and her saliva was collected. Now the rank smell of stinking fish became evident to everyone. Analysis of her urine revealed the presence of trimethylamine and a diagnosis of fish odor syndrome was finally made. She was instructed to minimize her intake of fish, liver, meat, eggs, and legumes. Her depression disappeared along with the odor of rotten fish. For the first time in years, our fish odor syndrome victim woke up in the morning without breath that would make a thousand cats drool.

HIGH ON PROTEIN

At what point am I allowed to say, "I told you so"? That's the question the late Dr. Robert Atkins asked in a piece he wrote for *Time* magazine in September 2003. In the article he described his thirty years as the lone voice in the wilderness crying out in favor of low-carbohydrate diets in face of formidable opposition by the "establishment," which labeled such an approach unscientific and dangerous. But the tide was turning, he maintained. Research was on the verge of vindicating his method. Indeed, several recently published papers have demonstrated that the Atkins diet, over the short term, can lead to greater weight loss than a low-fat diet and that, surprisingly, this high-fat, high-protein approach is not accompanied by unfavorable changes in blood cholesterol and triglyceride levels. As a result, a number of newspaper columns with alluring titles like "What If It Has All Been a Big, Fat Lie?" have appeared, presenting Atkins as the people's champion who, after a long struggle, finally succeeded in getting the grudging approval of the mainstream medical community. Still, maybe we should think twice about gorging on lobster with butter sauce or drowning our sorrows in as many bunless bacon cheeseburgers as our hearts desire!

Dr. Atkins, who was trained as a cardiologist before turning to the lucrative field of obesity treatment, was not the first to come up with the idea of a low-carbohydrate diet. That honor goes to William Banting, a nineteenth-century coffin maker. When he was in his thirties, Banting started to put on weight and sought medical help. He was subjected to various dietary schemes and was told to row every morning, take Turkish baths, and swallow concoctions of unknown origins. Nothing worked. The pounds just kept piling on. He was only 5 feet, 5 inches tall, but by the age of sixty-four he weighed over 200 pounds.

Tying his shoes was a real problem and he had to walk down stairs backward to reduce the stress on his knees. But it was when his hearing and eyesight began to fail that Mr. Banting really panicked. He consulted Dr. William Harvey, a famed ear, nose, and throat specialist, who was actually more interested in Banting's weight problem than in his ears or eyes. That's because Harvey had recently become captivated by the work of French physiologist Claude Bernard, who was studying the way nutrients affected the body. For reasons that aren't clear, Dr. Harvey became convinced that starches and sweet substances formed fat in the body and advised Banting to give up bread, butter (which he mistakenly thought contained starch), milk, sugar, beer, and potatoes. Lo and behold, the fat started to melt away, and within a year Banting had lost 46 pounds. Curiously, his hearing had also been restored. The public had to hear about this approach, Banting concluded.

In 1863, William Banting published a small booklet titled "Letter on Corpulence Addressed to the Public," which detailed his diet plan and his miraculous conquest of obesity. The medical establishment protested, and the "Banting diet" was ridiculed for being unscientific. In fact, the attacks were remarkably similar to those launched against Robert Atkins when he resuscitated this dietary approach in the 1960s. But Atkins had medical credentials, and from the beginning he successfully played the role of a maverick who dared to confront the dogma that counting calories is essential for losing weight. The growing number of Americans with expanding waistlines eagerly swallowed the Atkins plan. Certainly, the notion that you could lose weight while eating as much steak and eggs as you could stomach was an appealing one. The idea was also financially appealing to Atkins, who made a fortune from his best-selling diet books.

Capitalizing on his growing fame as a knight in shining armor battling the dragon of the "medical establishment," Atkins launched the Atkins Center, which promoted a number of "alternative" treatments. Ozone therapy for AIDS, ultraviolet blood irradiation for cancer, and a plethora of dietary supplements for rebuilding the immune system appealed to patients who were suspicious of mainstream medical care. Indeed, that most of Atkins' ideas were shunned by leading experts, coupled with his allegation that he was being persecuted, seemed only to increase his appeal. Still, it was in the area of weight management that Atkins' fame soared.

He claimed to have successfully treated thousands upon thousands of obese patients, yet, strangely, he never published these results. Surely the *New England Journal of Medicine* or the *Journal of the American Medical Association* would have been thrilled to publish such spectacular findings. Such a publication would also have legitimized Atkins' work and silenced his critics. Others did take up the challenge of investigating the low-carbohydrate approach scientifically. The May 2003 issue of the *New England Journal of Medicine* reported on two such studies. In one conducted over six months, severely obese subjects did lose more weight on a low-carb diet than they did on a low-fat diet, but the results (an average 5.8-kilogram loss) were hardly spectacular and the dropout rate was dreadful. In the other, there was a difference between the diets at six months, but not after one year. But the most telling analysis was published in the *Journal of the American Medical Association* by researchers who examined all the studies published on low-carb diets since 1966 and concluded that there is insufficient evidence to make a recommendation for or against these regimens, particularly over the long term. They found that when the low-carb diets worked, it was because they actually turned out to be low-calorie diets; in fact, weight loss was not associated with reduced carbohydrate content.

To me, whether the Atkins diet is more effective than what experts have suggested is a moot point. Health is a matter of more than just body weight. Knowing what we do today about the benefits of fruit consumption, recommending a diet that is so low in fruits is irresponsible. Scientific research has also linked excessive meat consumption to colon and prostate cancer and high protein intake with calcium loss from bones. Indeed, it will be interesting to compare disease patterns in people who have followed Atkins over the long term with those who have followed other diets.

Can we say anything in favor of Atkins? Yes. His attacks on intake of refined sugar and foods that readily elevate blood glucose are well justified. We consume far too much sugar, particularly in soft drinks. But you don't have to follow the Atkins plan to cut down on sugar. A diet high in whole grains, fruits, and vegetables, with small servings of meat and minimal sweets and soft drinks, is conducive to weight loss and good health. So I have a simple answer to Dr. Atkins' question about when he would have been able to say, "I told you so": Never!

MAD HONEY

Emergency rooms in hospitals along the West Coast sometimes post signs reminding physicians to consider honey poisoning as a possibility in the differential diagnosis of an apparent heart attack. How can honey, which in many people's minds is the prototypical "health food," poison anybody? Well, it can if the bees collect their nectar from flowers of the rhododendron family. These attractive flowers contain natural toxins known as "grayanotoxins," which can cause weakness, a slow heart beat, perspiration, and nausea—precisely the symptoms that can make a physician suspect a heart attack.

Before we take a look at what grayanotoxins actually do, return with me for a moment to 400 B.C., when Persian King Artaxerxes II defeated an army of Greek mercenaries led by his brother Cyrus, who had tried to wrest from him his throne. Cyrus was killed in the battle and the Greeks retreated under the command of the Athenian general, Xenophon. On their homeward journey they pitched camp at Trabzon on the Black Sea, an idyllic site. The hills were covered with rhododendrons and the woods harbored numerous beehives. The hungry soldiers feasted on the honeycombs and the honey with some dramatic results. Xenophon's journal tells the story: "All the soldiers who ate of the honeycombs lost their senses, and were seized with vomiting and purging, none of them being able to stand on their legs. Such as had eaten much more were like mad men and some like persons on the point of death." Xenophon also tells us that the "mad honey" did not have a long lasting effect: his soldiers recovered within twenty-four hours.

The Greek general's careful recording of the bizarre honey affair led to what may well be the first example of biological warfare in history. In 67 B.C., the Roman General Pompey attacked King Mithidrates of Pontus, whose armies had to retreat in the face of the superior Roman forces. But Mithidrates had an ace up his sleeve in the person of his physician, Kateuas. The good doctor was familiar with the writings of Xenophon and advised Mithidrates to retreat toward Trabzon and take advantage of the potential biological trap the countryside offered. Indeed, the Romans, like the Greek mercenaries three centuries before, decided that Trabzon was an ideal place for a campsite. History repeated itself and the Roman soldiers feasted on the treasure of beehives. The grayanotoxins struck again, and so did Mithidrates. His army massacred the Romans, who had been weakened after falling prey to Trabzon's "mad honey."

The mechanism of action of grayanotoxins has been fairly well established. These molecules bind reversibly to cell membranes and facilitate the passage of sodium across the membranes. Since normal cell function depends on the carefully regulated diffusion of ions such as calcium, potassium, and sodium across these membranes, any alteration in the transport system can have serious consequences. Luckily, the grayanotoxin binding is not permanent, so the effects on nerve and muscle cells wear off relatively quickly. Today, hospitals in Trabzon (which is in Turkey), still report a number of cases of mad-honey poisoning annually. Complaints usually start an hour after a person eats at least 50 grams of honey. The slow heart beat, impaired breathing, nausea, and vomiting are much like those symptoms experienced by Xenophon's men. Sometimes atropine is needed to boost the heart rate and in rare cases a pacemaker has to be temporarily installed to correct an irregular heartbeat.

As the signs in the emergency rooms of the Pacific Northwest make clear, you don't have to travel to Trabzon to fall victim to mad honey. Wherever rhododendron species such as western azalea are common, honey poisoning is a possibility. Actually, you don't even have to be in the Pacific Northwest. In the Appalachians, mountain people have a long tradition of feeding wild honey to their dogs before eating it themselves. If the dog shows no ill effects, the feast is on!

Before you become concerned about eating honey, it is important to realize that the product of commerce is pooled from many areas and any grayanotoxins present are effectively diluted. Most poisonings can be traced to producers who have only a few hives. So the average consumer's chance of having a mad-honey experience is pretty remote. But that doesn't mean honey is in the clear. In fact, believe it or not, probably no other food contains as many potential toxins as honey. And we're not talking about synthetic toxins, although honey can certainly contain those as well. Those busy little bees gather nectar from a huge variety of plants, many of which harbor a variety of natural toxins. These are not necessarily the kind of toxins that will lay you low quickly, like the grayanotoxins; they are more insidious, possibly causing long-term health problems.

About 3 percent of all plants in the world produce compounds called "pyrrolizidine alkaloids," probably to protect themselves from insects and fungi. In rats these compounds have been shown to cause cancer, and any substance that produces such an effect obviously alarms us. In fact, Germany has established regulations for herbal products containing these alkaloids, making 0.1 micrograms per day the maximum permissible intake. Products for pregnant or lactating women cannot contain any of these alkaloids at all. Bees collect nectar from all sorts of plants, including those that manufacture the pyrrolizidine alkaloids. Indeed, if the German regulations for herbal products were applied to honey, many varieties could not be sold. In a perfect world, all honeys would be analyzed for pyrrolizidine alkaloids and those containing excessive amounts would be taken off the market. But we don't live in a perfect world. And that is just the point! There are numerous natural toxins out there conspiring to undermine our health without arousing great concern, simply because people don't know that they exist. Activists organize demonstrations against well-tested products of biotechnology, but I have yet to see them walk down the street with banners protesting the presence of grayanotoxins or pyrrolizidine alkaloids in our food supply. Scientifically, they would have a much better case for the latter.

When Sugar Isn't So Sweet

How would you like to be on a diet that allows no sugary desserts, no carrots, no tomatoes, no corn, no cold cuts, no hot dogs, no sweetened cereals, and no fruit? And not just for a little while, but for the rest of your life. Furthermore, you had better stick to the diet or there will be no rest of your life. Luckily, only about 1 in 20,000 of us is forced to follow this bizarre dietary regimen: the one who is unfortunate enough to be born with hereditary fructose intolerance (HFI). The one made sick by just the thought of anything sweet.

Let's get something straight right away: HFI is a completely different condition from dietary fructose intolerance, also known as fructose malabsorption. If you've been told by your physician that the latter is what you have, you need not fear for your life. You may experience some unpleasant diarrhea, pain in the gut, or the production of impressive amounts of gas, but you will survive! Fructose malabsorption occurs when, for some unknown reason, fructose is not absorbed from the gut into the bloodstream and passes through to the colon where bacteria readily digest it. This bacterial digestion produces gas as well as compounds that trigger diarrhea. While HFI and fructose malabsorption are different diseases, the treatment for each is the same: avoiding fructose!

Fructose is a sugar. When we use that word in everyday language, we generally refer to the white crystals we put in our coffee or tea. To the chemist, however, *sugar* refers to any one of a number of water-soluble substances that have a certain commonality in molecular structure and serve as the body's prime source of energy. The scientific term for table sugar is *sucrose*. To complicate things further, sucrose is actually a disaccharide made of glucose and fructose, two smaller sugars that are bonded together. Starches are composed of long chains of glucose molecules.

In the digestive tract, starch is degraded to glucose and sucrose is broken down into glucose and fructose. The body can use both glucose and fructose as sources of fuel. To put it simply, our bodies burn simple sugars for energy, just as a car burns gasoline. In both cases the major products of the combustion process are water and carbon dioxide. But the body's conversion of simple sugars into carbon dioxide and water is a complicated business that requires many steps and biological catalysts known as "enzymes."

The unfortunate people who suffer from HFI lack aldolase B (also called "fructose-1-phosphate aldolase"), one of the enzymes needed to metabolize fructose. This lack results in an accumulation of fructose-1-phosphate in the liver and kidneys, potentially causing these organs to fail. Aldolase B is also produced in the healthy intestine, and a deficiency there can cause severe abdominal pains. But accumulation of this substance has another consequence: it interferes with the body's use of glucose for energy. Glucose is normally stored in the liver in the form of glycogen, a polymer that, like starch, is made of a long chain of glucose molecules. Glycogen releases glucose as needed for energy production. A buildup of fructose-1-phosphate blocks the breakdown of glycogen and results in hypoglycemia, or low blood sugar. The result is that affected people experience symptoms of nausea, vomiting, sweating, tremors, confusion, and even seizures following the ingestion of any food that contains fructose.

Diagnosing hereditary fructose intolerance is not easy. Symptoms generally appear when a baby is weaned and begins to consume foods that contain fructose. But there are many other diseases that may be accompanied by vomiting, dehydration, and liver problems, which are typical signs of HFI. A liver biopsy to test for the missing enzyme or an in-hospital challenge with fructose can confirm the diagnosis, but both these procedures are somewhat risky. If the condition goes undiagnosed, however, death from liver or kidney failure can result. On the other hand, if the correct diagnosis is made, a diet that strictly eliminates fructose leads to a complete recovery. Then there are the cases that fall in between death and recovery. These are the people who survive HFI through infancy but lead miserable lives, suffering from symptoms that are not recognized as being associated with HFI.

A fascinating example of an in-between case is a young Englishman who had a history of infrequent seizures over a ten-year period and went through the usual battery of tests aimed at detecting brain abnormalities. His doctors could find nothing wrong. The only information the patient was able to provide was that he had always intensely disliked sweets, an observation his physicians were unable to relate to his medical problems.

The mystery was finally solved by the patient himself. He happened to read a newspaper article about HFI and recognized the symptoms as his own. Upon reflection, he realized that his attacks of weakness and seizures always followed the ingestion of sugar. Tests revealed that he did indeed suffer from the disease. The required treatment—avoiding fructose—was simple, at least in theory. But this is not such a simple solution in practice. Since table sugar, or sucrose, breaks down in the body into glucose and fructose, it must be eliminated. Fruits and many vegetables contain fructose and therefore must be avoided. Even sorbitol, a sweetener found in many "dietetic" products, has to be shunned because the body converts it to fructose. Rhubarb, asparagus, cauliflower, and spinach are fine, but ketchup, honey, fruit juices, and, of course, sugary soft drinks are no-nos. Glucose can be used as a sweetener, and, in fact, the hypoglycemic seizures characteristic of fructose intolerance can be treated with it. Our English patient never leaves home without his glucose tablets, just in case he should accidentally ingest some fructose.

Clearly, an understanding of the chemistry of fructose metabolism has led to at least at partial solution to the problem of hereditary fructose intolerance. But it is unfortunate that just as modern research is demonstrating the beneficial effects of eating fruits and vegetables on cancer and heart disease, the sufferers of HFI have to be counseled to stay away from these foods. Obviously, one person's treat is another's poison! Can the fructose-intolerant people take solace in anything? Why, yes! Since they avoid sugar, they hardly ever get cavities. Presently, the only way to prevent HFI is through DNA analysis of prospective parents. If both parents carry the gene, advice about family planning can be sought. And for those of us who have gone through life enjoying all fruits and vegetables and the occasional sweet dessert, we can be thankful that we lucked out with sweet genes.

SUGAR-FREE KISSES

The sign above the Hershey's chocolate store in New York's Times Square is truly awesome. Actually, I should say "signs." As you gaze up at this fifteen-story spectacle, you can't help but be impressed by the smoking chimneys of the whimsical chocolate factory, the steaming cup of cocoa, the beaming pyramid of Hershey's Kisses, and, of course, the giant chocolate bar. Can anyone resist going inside? I sure couldn't. I was immediately welcomed by a giant dancing Hershey's Kiss, and as I scampered aside to avoid being bussed I ran smack into a huge display of Hershey's new sugar-free chocolates. "Aha," I thought. "There's a story in here somewhere!" And there was.

Why should anyone be enticed by the prospect of sugar-free chocolates? Obviously they appeal to diabetics and consumers worried about cavities. But as I eavesdropped on the conversations of the chocoholics who were snapping up the sugar-free bars like candy, it became apparent that many thought the Holy Grail of chocolate manufacturing had finally been discovered. They thought they were buying bars that, because of the absence of sugar, were low-calorie products and therefore "healthier." They were mistaken.

As far as sweetening power goes, the sugar in chocolate can be replaced by artificial sweeteners such as aspartame, acesulfame-K, or sucralose. All of these substances are hundreds of times sweeter than sugar, which means that only small amounts are necessary. The tiny quantity, though, presents a problem. Sugar not only sweetens but also provides bulk, texture, and an appealing feel to chocolate. You cannot make an appealing bar by simply replacing sugar with artificial sweeteners. Enter the sugar alcohols, or, as they are commonly known, the "polyols." Polyols are carbohydrates that provide sweetness but are metabolized by the body in a different way than sugar. They are manufactured from naturally occurring sugars by a straightforward chemical reaction. Lactitol, for example, the polyol used in the Hershey products, is made by reacting the milk sugar lactose with hydrogen gas. Similarly, glucose can be converted to sorbitol, maltose to maltitol, and mannose to mannitol—all these polyols are used in a variety of sugar-free gums, ice creams, candies, and cookies. Polyols are effective sugar replacements because they can more or less be substituted for an equal amount of sugar. However, since they are somewhat less sweet than sugar, an artificial sweetener such as sucralose is also added to boost sweetness. But what is the point of replacing one carbohydrate with another?

Sucrose, or table sugar, is composed of a molecule of glucose joined to a molecule of fructose. During digestion in the stomach and small intestine, this link is broken and the glucose and fructose are absorbed into the bloodstream, ready to serve as a source of energy. One gram of sucrose contains four calories, which means that we have to spend four calories' worth of exercise to use up the sugar. If we don't, the excess sugar is converted into fat, ready to be stored by the body. Now let's look at the polyol lactitol. This compound resists being absorbed into the bloodstream from the stomach and small intestine. While some is slowly absorbed, much of the lactitol moves down through the small intestine and migrates into the colon. Here it encounters a variety of bacteria that call this region of our digestive tract home. Some of these microbes consider lactitol a tasty morsel and make a meal of it. Unfortunately, these bacteria are quite flatulent and produce gases as they dine on the lactitol. Buildup of the effluvia can cause bloating and cramping. Furthermore, the body tries to eliminate unabsorbed lactitol, sometimes resulting in an unpleasant laxative effect. So, what's the upside to polyols?

First of all, a nutrient that is not absorbed by the body cannot contribute calories. Lactitol, which is only partially absorbed, provides two calories per gram compared with the four supplied by the same amount of sugar. Basically this means that only half as much activity is required to "burn up" the calories in 1 gram of lactitol compared with 1 gram of sugar. Remember, though, that most of the calories in a chocolate bar come not from the sugar but from the fat in the cocoa butter used to make the chocolate. And sugar-free chocolates contain no less fat than regular chocolates. Replacing sugar with lactitol results in only about a 20 percent savings in calories—not a particularly significant amount. An interesting potential benefit, though, is lactitol's ability to serve as a "prebiotic." At a daily dose of 5 to 10 grams it encourages the growth of beneficial bacteria in the colon at the expense of disease-causing bacteria. Some of the organic acids, metabolites of the beneficial bacteria, have potential anticancer properties. Then there is the fact that while bacteria in our colon like lactitol, those in our mouth do not; therefore, they do not churn out the cavity-causing acids they would if fed sugar.

Now, what about the portion of lactitol that is absorbed into the bloodstream? Unlike most carbohydrates, it is not readily converted to glucose and therefore is less likely to trigger an insulin response. As a result, diabetics who have to count carbohydrate exchanges can eat more of the sugar-free chocolate than regular chocolate for the same exchange value. Whether or not one wants to eat more of this chocolate is another question. The party line is that lactitol and other sugar alcohols, when consumed in moderation, should produce no undesirable side effects. But the reality is that in some people even small doses of sugar alcohols can cause temporary bloating, diarrhea, and impressive flatulence. Of course, I had to try the sugar-free Hershey's chocolate. My digestive tract didn't rebel, but my taste buds were less than enthralled. Give me a small piece of Lindt milk chocolate any day. I was far more impressed by the giant Hershey bar outside the building than by the edible bar inside.

FOR THE LOVE OF BROCCOLI

Paul Talalay eats his sprouts. He not only eats them but also sells them. And tea made from them. But you won't find Talalay behind the counter in some health food store. In fact, he scorns many of the overhyped, overpriced, underresearched products with which these stores entice customers. Where you will find the sprightly eighty-year-old is in the hallowed halls of Johns Hopkins University in Baltimore, Maryland, where for many years he was the director of the School of Medicine's Department of Pharmacology and Experimental Therapeutics and is now the John Jacob Abel Distinguished Service Professor of Pharmacology. Just mention Dr. Talalay's name in scientific circles and the conversation will immediately shift to "chemo-protection" and of all things, broccoli!

Talalay's fifty-year research career has focused on the prevention and treatment of cancer. As a young medical student he was intrigued by the case of a prostate cancer patient who responded dramatically to therapy with steroids. Were there other substances that could also affect this dreaded disease in a similar fashion and maybe even prevent it? Talalay decided to devote his career to finding out. In 1992 he finally made a discovery that would not only tantalize the cancer research community, but would also splash his name across the pages of newspapers. Researchers had long known that populations who ate lots of vegetables had lower rates of several types of cancer. But why? Was it some specific compound or set of compounds found in these foods that was responsible? Talalay seemed to have found an answer.

He had isolated a compound called "sulforaphane" from broccoli, which at least in laboratory experiments had demonstrated anticancer properties. In mouse cells grown in tissue cultures, sulforaphane boosted the production of so-called Phase II enzymes. These enzymes form part of the body's protection system against foreign intruders, including carcinogens. Glutathione-S-transferase, for example, binds to carcinogens and removes them from the body. The body regards sulforaphane as a foreign substance and cells gear up their biochemical machinery to produce Phase II enzymes to eliminate it. The enzymes then remove sulforaphane as well as many other foreign substances they encounter.

Inducing protective enzyme formation in cell cultures is one thing; protecting live animals from cancer is quite another. For Dr. Talalay, the obvious next step was to treat rats with sulforaphane before attempting to induce tumors in them with a known carcinogen. He used dimethyl benzanthracene, a potent inducer of breast tumors, and the results were astounding! Almost 70 percent of the control rats developed cancer while tumors were detected in only 35 percent of the rats that had been dosed up on sulforaphane. Other experiments showed that sulforaphane also offered protection against colon cancer, a type of cancer that has been linked to carcinogens found in foods such as barbecued meat. But what did this finding mean for humans? After all, the rats' diet was not nearly as varied as humans'. Furthermore, the amount of sulforaphane that offered the rats protection against cancer corresponded to humans' eating several pounds of broccoli per week.

Two possibilities presented themselves to Dr. Talalay: find a better source of dietary sulforaphane or investigate the use of isolated sulforaphane supplements. The first option seemed more appealing because the nutritional literature is filled with examples of substances that perform quite differently when they are introduced in a pure form as opposed to being a component of food. Also, foods such as broccoli contain a number of other beneficial nutrients such as selenium, calcium, folic acid, and vitamin K. It was at this point that Dr. Talalay learned that broccoli sprouts could potentially yield as much as fifty times more sulforaphane than adult broccoli. Why potentially? Because neither broccoli nor its sprouts actually contains sulforaphane; what each has is glucoraphanin, a compound that yields sulforaphane when it reacts with the enzyme myrosinase. This enzyme is liberated when the plant's tissues are disturbed by chopping or chewing. Cooking destroys the enzyme, but fret not because bacteria present in our gut can also break down glucoraphanin to yield sulforaphane.

Talalay and his coworkers studied various broccoli varieties and through a laborious process selected seeds with the highest content of glucoraphanin. So convinced were they of the potential nutritional benefits of sprouts grown from these seeds that Talalay and plant physiologist Jed Fahey launched Brassica Protection Products, a company that would market BroccoSprouts with part of the profits going to research into cancer chemoprotection. These sprouts are guaranteed to yield twenty times as much sulforaphane as mature broccoli. To be sure, the benefits of sulforaphane so far have been shown only in cell cultures or in animals. Dr. Talalay would be the first to agree that reducing the risk of cancer takes more than just eating BroccoSprouts and that human trials are sorely needed. He has already begun to investigate if Phase II enzymes can be elevated in humans with BroccoSprouts and envisions trials in high-risk populations such as people with a family history of breast cancer or a history of colon polyps.

Commercializing the broccoli sprouts led to another amazing discovery: employees at the sprouting facilities took to snacking on the sprouts they were producing. A couple that had been suffering from stomach ulcers for a long time claimed that the sprouts cured them. This claim did not come as a complete surprise, because earlier experiments had shown that broccoli possessed some antibiotic properties and the link between ulcers and infection with the bacteria *Helicobacter pylori* is well established. Test-tube studies quickly showed that purified sulforaphane killed forty-eight different strains of the bacteria. This is an exciting finding because *Helicobacter* infection is also a risk factor for stomach cancer. Preliminary studies have already shown that sulforaphane can reduce stomach tumors in mice—and at a dose that does not translate to a human having to eat mountains of broccoli sprouts. A daily snack is all that is involved.

The Japanese are an ideal population to investigate in this regard because about 80 percent of adults are infected with *Helicobacter* and the incidence of gastric cancer is extremely high. A clinical trial is now being mounted in Japan to investigate the effects of BroccoSprouts on both *Helicobacter* and the incidence of cancer. Wouldn't it be great if something as simple as snacking on the right kind of broccoli sprouts could indeed reduce the risk of ulcers and stomach cancer? Anyway, you now know why Paul Talalay eats his broccoli sprouts regularly. Maybe we should too.

I'M A BREAD AND PROPIONATES MAN

I'm a bread man. Have to have it for every meal. Well, maybe not *every* meal. Pizza without bread on the side may be OK, but that's about the only exception. I am particular, though, about the kind of bread I eat. I am not the least bit enticed by the spongy, tasteless, pale, packaged loaf that masquerades as bread in many a supermarket, but give me a slice of Austrian Schinkenbrot smeared with a bit of butter and I'm in heaven. That's right, butter. In moderate amounts it can fit into a healthy diet. And to me, it sure tastes better than margarine.

Certainly Schinkenbrot is not one of your more common bakery items. So I really wasn't surprised to see a fellow shopper eyeing the sample in my cart with some suspicion. The bread quickly proved a catalyst for conversation and I explained that I was partial to this loaf not only for its flavor, but also for its nutritional value. It's basically made of whole rye kernels, whole wheat flour, rolled oats, and sour dough culture, ingredients that are surely superior to the refined white flour and sugar that dominate the classic American "toast" bread. Intrigued, she picked up my Schinkenbrot and began to peruse the ingredients. I saw her brow furl as she read the list, and with an air of incredulity, she lifted her eyes and pierced me with a look that I thought would have been reserved for someone who had committed a capital crime. "You're not really going to eat this, are you?" she sputtered with considerable bewilderment. After confirming that I had not been planning a fun-filled afternoon of pigeon feeding in the park and the bread was indeed destined for the dinner table, I queried her concern about my diet. Her reply? "It says right here, 'May contain calcium propionate!'" She blurted out the term "calcium propionate" as if it were synonymous with "poison," which alerted me to what this interaction was all about: another rampant case of chemophobia.

I thanked the shopper for bringing this issue to my attention, because I had never noticed the word "may" on the label of my preferred bread. I don't like the notion of "may contain." I would prefer a guarantee that the bread does contain calcium propionate. You see, I'm not a great fan of moldy food. Molds not only make for unsightly green splotches, but some produce decidedly dangerous compounds, which is why we add preservatives such as calcium propionate to foods. It prevents the growth of molds while allowing yeast to flourish—an obvious advantage when it comes to baking bread. And that's not all that calcium propionate does. It also inhibits the formation of "rope" in bread. The spores of certain bacteria, such as *Bacillus mesentericus,* are often present in flour and germinate under the moist, warm conditions needed to make bread rise. These bacteria are not harmful to humans, but they change the texture of the dough and produce sticky yellow stringy patches that make for an unpalatable bread. Propionates prevent this undesirable result.

Are propionates safe to eat? Sure they are. Food producers can't just randomly add chemicals to their products. Additives are strictly regulated and must show a clearly demonstrated benefit with minimal risk before they are allowed into general use. In the case of propionates, safety is not hard to demonstrate. These compounds cruise through our bodies all the time, and they don't have to be introduced through bread. Bacteria in our intestine feed on fiber, the indigestible part of fruits, vegetables, and grains, and convert it into a variety of compounds that include propionic acid. This acid is then absorbed into the bloodstream. Far from being harmful, some studies have shown that such short-chain fatty acids can reduce the risk of colon cancer and may even be helpful in preventing other diseases of the digestive tract.

Propionates, as derivatives of propionic acid are called, also occur naturally in our food supply. Perhaps the best example is Swiss cheese. The texture and flavor of this cheese is due to the addition of a starter culture that includes the bacterial species known as *Propionibacter shermanii.* These bacteria break down some of the fat to produce carbon dioxide gas, which explains the presence of holes in the cheese. They also produce propionic acid, which is responsible for some of the cheese's characteristic nutty flavor. Swiss cheese contains roughly 1 percent propionates by weight, far more than the amount used as a preservative in bread.

So I am absolutely untroubled by the presence of calcium propionate in my Schinkenbrot. Nor will I be put off my bread by the carbophobes who believe that the key to health, and weight control, lies in curtailing most carbohydrates. Actually, a recent study in Finland showed that elderly men who ate just three slices of old-fashioned, fiber-rich rye bread per day reduced their risk of fatal heart attacks by 17 percent. (We're not talking about the type of rye bread sold in most bakeries here; that version is made mostly from refined flour.) And if you are interested in weight control, you may find comfort in an Australian study for which researchers fed volunteers seven different kinds of bread and rated their "satiety" value after two hours. Soft white bread scored the lowest. Coarse-textured, high-fiber breads ranked the highest. Subjects who ate the latter consumed fewer calories over the rest of the day.

Is there any problem with eating high-fiber dark breads like my Schinkenbrot? Maybe a little one. These dark breads are often referred to as "pumpernickel," from the German words *pumpern*, meaning "to break wind," and *nickel*, meaning "devil." Pumpernickel was thought to be so hard to digest that its victims would pass wind like the devil. I am not privy to Lucifer's dietary habits, so I do not really know what it means to pass wind like the devil. But it may not be a bad thing. Passing gas is a sign of high fiber intake, which has all sorts of health benefits. Perhaps I should have mentioned this fact to my grocery-aisle acquaintance when I encountered her again in the checkout line. She could have used the info because I noted a loaf of "organic" white bread in her carriage, whose package proudly declared,

"No preservatives added." Oh well. Maybe she likes mold. Or maybe she just prefers days with no wind.

PUT GARLIC WHERE?

Just about fifty years ago, Adolphus Hohensee glanced around the room and told the members of his audience that they were obviously vitamin-deficient, mineral-starved, cooked-food-enervated, constipation-befouled, oxygen-deprived, sugar-acidified, meat-polluted, starch-clogged, and gravy-saturated—just like all Americans. This self-proclaimed nutritional expert, who actually had no scientific background at all, then went on to describe how these conditions could be remedied. Garlic, he maintained, would cleanse the intestines, purify the blood, and restore health. He advised the nightly insertion of a clove into the rectum. Garlic breath in the morning would be proof that the clove had worked its magic and cleansed the system from bottom to top.

Of course, Hohensee was not the first to ascribe miraculous healing properties to garlic. The ancient Egyptians fed large doses of garlic to their slaves to keep them strong and healthy; the Greeks claimed that garlic would "open obstructions" in the body; and Indians use the bulb to make lotions that have long been used in the East for washing wounds and ulcers. But Hohensee's exhortations coincided with the beginning of the current enthusiasm for the use of various garlic preparations. Sales amount to some $70 million per year in North America.

Before examining the scientific validity of this enthusiasm, let's take a quick look at the chemistry of garlic. It isn't simple! Fresh garlic has virtually no smell, and only when the bulb is sliced or crushed does the "stinking rose" begin to live up to its reputation. That's when the disrupted cells release the enzyme alliinase, which converts odorless alliin to odiferous allicin. Allicin is unstable and immediately begins to break down, eventually yielding a number of other compounds that are responsible for whatever health benefits garlic may have. Since allicin is the parent compound of this garlic cascade, it is the one often touted on the label of supplements. But any claim that a product contains allicin can be dismissed out of hand, as the compound is far too unstable. Claims about "allicin yield" or "allicin potential" are somewhat more meaningful, but not by much. Manufacturers usually determine yield by mixing crushed tablets with water and measuring the amount of allicin released. But this model is not an appropriate one for what happens in the body.

Realistically, any garlic supplement has to be protected from contact with stomach acid, which immediately destroys alliinase and makes the release of allicin impossible. This protection is usually achieved by encapsulating the contents of the supplement in gelatin or coating the pill with cellulose or polyacryclic acid derivatives that dissolve only in the less acidic conditions of the intestine. A fitting test for allicin release is one sanctioned by us Pharmacopeia (method 724A) that simulates the conditions encountered by a pill as it travels through the digestive tract. When this test is applied to commercial garlic supplements, the results are astounding. Over 80 percent of products tested release less than 15 percent of their claimed allicin potential! Clearly these tablets do not deliver the allicin dosage that is thought to be therapeutic. Consequently, we have to struggle with the question of whether garlic at any dose is therapeutic.

This question can only be answered by human trials. It's nice to know that some garlic extract retards cholesterol oxidation in cells or may reduce carcinogen formation in a test tube. However, these tests do not mean these same things happen in the body. Numerous studies of garlic's effects on health have been carried out. Early studies suggested a cholesterol-lowering effect and received widespread publicity. Unfortunately, more sophisticated and better studies curtailed the initial optimism. Perhaps the best way to evaluate studies that show contradictory results is to perform a "meta-analysis," or pool the results and analyze them together. When researchers isolated the best garlic studies and combined the results, they found, much to their disappointment, that garlic's ability to reduce cholesterol was minimal. The effect on blood pressure was insignificant. Still, companies keep promoting their supplements based on the early, poor-quality studies. One German manufacturer still touts its garlic supplement in North America as "clinically proven to reduce cholesterol," ignoring that five out of six of the studies it has sponsored since 1995 have not shown a cholesterol-reducing effect. The German government, though, finds the studies convincing and no longer allows cholesterol-lowering claims on behalf of garlic. And this is a government that is traditionally very friendly to herbal remedies. While garlic may not reduce cholesterol, it may still have an effect on heart disease. Ajoene, one of the breakdown products of allicin, may reduce the risk of heart attacks by preventing the formation of blood clots.

The garlic situation is more encouraging with respect to cancer, perhaps because most studies investigated the effect of raw or cooked garlic instead of supplements. A meta-analysis showed that consuming an average of six or more cloves per week lowered the risk of colorectal cancer by 30 percent and stomach cancer by 50 percent when compared with the consumption of less than one clove per week. Even the risk of prostate cancer may be reduced. A US National Cancer Institute study of men in Shanghai showed that eating a clove per day reduced the risk of this cancer by over 50 percent. There is a caveat to these types of studies, however. Consumption of specific foods is determined by means of questionnaires, and peoples' memories may not be all that reliable. Furthermore, heavy garlic consumption may just be the hallmark of a mostly vegetarian diet.

It seems then that while it is difficult to make a case for taking garlic supplements, incorporating as much garlic into one's diet as one's social life will allow is a good idea. There's no doubt, though, that garlic odor can cause discord. An Atlanta psychoanalyst once went so far as to prescribe garlic to a woman whose boyfriend threatened to commit suicide if she broke up with him. The psychoanalyst suggested that she start eating garlic and rub a clove on her arms and neck before a date. After just two of these spicy encounters the boyfriend had had enough! And what can you do if you love both your mate and garlic? Parsley oil has a reputation for neutralizing garlic aroma, so chewing a few sprigs of parsley after consuming garlic may help. Or not. Adolphus Hohensee would have liked that idea because parsley is a good source of chlorophyll, a substance he maintained was a cure for cancer. This dreaded disease, he claimed, was caused by eating unwholesome processed foods. It seems, however, that Hohensee suffered from a bit of the "do as I say and not as I do" syndrome. His star began to plummet when he was caught by a photographer dining on fried snapper, French bread, apple pie, and beer. And this just after he had finished giving a lecture that castigated these foods! History does not record whether the nutritional guru had taken precautions against these dietary villains by self-administering garlic in his peculiar fashion.

DIETARY SUPPLEMENTS:
TO TAKE OR NOT TO TAKE?

Should you take a daily multivitamin/mineral supplement? It's a pretty simple question to answer, one would think. After all, there have been literally thousands of studies on how vitamin and mineral intakes relate to health. More than 100 million people in the US and Canada believe the question has been answered and take a variety of daily supplements to protect themselves against disease. Can all these people be wrong? Let's see.

Last year, researchers in Oxford, England, decided to look into the issue of preventing heart disease through vitamin supplementation. Since it is well accepted that LDL cholesterol (the "bad" cholesterol in our blood) is more likely to accumulate in artery walls after undergoing reaction with oxygen, it was reasonable to expect that "antioxidants" such as vitamins E and C and beta-carotene would reduce the risk. Over 20,000 adults who had diabetes, high blood pressure, high cholesterol, or were otherwise at risk for heart disease, were enrolled in a major study. Half the subjects received a daily supplement of 600 international units (IU) vitamin E, 250 milligrams vitamin C, and 20 milligrams beta-carotene while the others got a placebo. Blood tests showed that this regimen significantly increased the levels of these nutrients in the blood. And what happened after five years of this regimen? Nothing. There was no difference in any form of disease, or in death rate, between the experimental group and the control group. Of course, one can argue that the subjects were already at high risk and the disease processes were already underway when supplementation was begun and the supplements were powerless to reverse them. So how about looking at giving supplements to an initially healthy population? That's been done too.

This study enlisted over 83,000 healthy American physicians who filled out questionnaires about their dietary habits. About 30 percent reported that they used daily multivitamin supplements. After five and a half years, just over 1,000 had died from some form of cardiovascular disease. There was no relationship to supplement intake. Doctors who had been popping vitamin pills were as likely to die from heart disease as those who had not. Now, one can argue that these physicians were cognizant of good nutrition, ate a balanced diet, and therefore had no vitamin deficiencies that could be corrected by taking supplements. And what about supplements and average people? A study on this question has also been completed.

When some 600 subjects over the age of sixty were assigned to take 200 IU vitamin E, a multivitamin, both, or a placebo, the results were surprising. Those who took vitamin E had worse colds and respiratory infections over a period of fifteen months. Worrisome? Then how about the study of over 70,000 post-menopausal nurses that showed those who consumed the most vitamin A from foods and supplements over an eighteen-year period had the greatest risk of bone fractures? Ready to flush your vitamins down the toilet? Well, hold on.

A study at Canada's Memorial University examined the effects of giving seniors a daily multivitamin/mineral supplement for a year. There was a significant improvement in the subjects' immediate memory, abstract thinking, and problem-solving abilities. An English study of over 20,000 people found that those who had the highest levels of vitamin C in the blood lived the longest. A large American study found that maternal use of vitamin supplements reduced the risk of neuroblastoma, a childhood cancer. It is well established that taking folic acid supplements during pregnancy can reduce neural-tube defects in babies. But that's not all folic acid can do. A statistical review of over 20,000 people in ninety-two studies showed a significant decrease in the risk of heart disease with higher blood levels of folic acid. And there's still more. Women who have higher levels of folate in their blood appear to have greater protection against breast cancer, especially if they consume two or more alcoholic beverages per day, a known risk factor for breast cancer. Reduced blood levels of folic acid are also linked to colorectal cancer and long-term use of supplements has been shown to lower risk by as much as 75 percent. Vitamin D intake also offers protection against breast cancer in some studies. Selenium deficiencies in the diet have been linked to prostate cancer. A randomized, double-blind, placebo-controlled study in North Carolina showed that respiratory tract infections could be dramatically reduced in diabetics who took a combination of vitamins and minerals. And here is something really intriguing: A study published in 2004 reported its findings after having followed almost 5,000 people aged sixty-five and older in Utah. Those who took daily supplements of vitamin E (400 iu) and vitamin C (500 milligrams) reduced their risk of developing Alzheimer's disease by almost 80 percent. Strangely, neither vitamin alone was effective. Ready to rescue those vitamins from the toilet?

The point of reviewing these studies is to show that it is possible to support either side of the supplement issue by selectively looking at the scientific literature. However, scientific conclusions should hinge on an examination of *all* the evidence. And that is just what researchers do when they establish the recommended daily doses of nutrients. Yes, it is possible to get such amounts from a balanced diet. But surveys show that while North Americans do not suffer from acute vitamin deficiencies, suboptimal intakes are common. About 75 percent of the population consumes inadequate amounts of folic acid. We know from laboratory experiments that low levels of folic acid lead to DNA damage and breaks in chromosomes, both of which are linked to cancer. Many people, especially seniors, fall short of the recommended daily intake of vitamin B_{12}, vitamin C, and zinc. The human body requires some forty micronutrients for effective function and while we may not be sure of the optimal amounts, we do know that a large segment of the population has inadequate intakes.

So I'm going out on a limb here, but I think it is a very sturdy limb. The benefits of taking a daily multivitamin/mineral supplement that has no more than 100 percent of the recommended daily intake of each nutrient greatly outweigh any risks. Look for supplement with no more than 4,000 IU vitamin A (better if some of this comes from beta-carotene, a vitamin A precursor); about 400 IU vitamin D; 90 milligrams vitamin C; at least 2.4 micrograms vitamin B_{12}; 400 micrograms folic acid; 10 milligrams zinc; and no more than 10 milligrams iron (although menstruating women may require up to 18 milligrams). The jury is still out on vitamin E, but up to 400 IU presents no problems. There is no consensus on vitamin C, but 500 milligrams per day is unlikely to cause any side effects. Some companies advertise that their products are "more pure" or "better absorbed." But all vitamins have to meet standards of purity and absorption. And there's no need for any ginseng, lutein, ginkgo, or whatever happens to be currently in vogue to be included. There will be no benefits from the amounts of these compounds put into a multivitamin. Above all, remember that Mae West's famous comment "Too much of a good thing is wonderful" most certainly does not apply to supplements. Her observation that "It takes two to get one into trouble" is far more appropriate.

What's In a Vitamin Name?

You've got to love the plot. A meteorite falls to earth and begins to ooze a revolting goo that dines on humans. It makes its way into a movie theater and, in a classic scene, surges from the projection booth, ready to gobble up everyone in its path. Audiences in 1958 were absolutely terrified by *The Blob,* which was destined to become a science-fiction classic. The film also introduced actor Steve McQueen to the world and made me a fan. I watched his films and followed his career until his unfortunate death from lung cancer in 1980. Steve introduced me to some great movies and also to vitamin B_{15}, which he had been taking at a Mexican cancer clinic in an attempt to halt the progress of his disease.

By 1980 I had been teaching for a while and had developed several lectures on vitamins. But none of the books or publications I consulted had ever referred to vitamin B_{15}. And for good reason, as it turned out. Vitamins are substances that must be included in the diet in order to maintain health and prevent certain deficiency diseases; they cannot be synthesized by the body. The first vitamin-deficiency disease to be recognized was scurvy, described as early as 1550 B.C. by the Egyptians in the *Ebers Papyrus*. In the sixteenth and seventeenth centuries, when long ocean voyages became common, thousands of sailors died from scurvy, which is characterized by spongy gums, loose teeth, and bleeding into the skin and mucous membranes. The first clue that scurvy was a diet-related disease came from North American Indians who showed French explorer Jacques Cartier that a brew made from pine needles could cure the condition. In 1747 James Lind, the Edinburgh-born naval surgeon, discovered that eating oranges and lemons prevented scurvy, but it took another fifty years before the British Navy required sailing vessels to carry supplies of lemons or limes (this requirement is the origin of the slang term "limeys" for British sailors). Around the same time, British Captain James Cook discovered that fresh fruits and sauerkraut also prevented scurvy. Finally, in the 1930s, Hungarian scientist Albert Szent Gyorgyi isolated the scurvy-protective factor and named it "vitamin C." Why? Because the idea of naming vitamins by letters had already been introduced some twenty years earlier and *A* and *B* were taken.

The letter designation for vitamins goes back to the early part of the twentieth century. When the mechanized rice mill was introduced in Asia, a new disease that came to be called "beriberi" appeared. *Beriberi* means "weakness" in the native language of Sri Lanka and describes a condition of progressive muscular degeneration, heart irregularities, and emaciation. Kanehiro Takaki, a Japanese medical officer, studied the high incidence of the disease among sailors in the Japanese navy from 1878 to 1883. He discovered that on a ship of 276 men, where the diet was mostly polished (i.e., hulled) rice, 169 cases of beriberi developed and 25 men died during a nine-month period. On another ship there were no deaths and only 14 cases of the disease. The difference was that the men on the second ship were fed more meat, milk, and vegetables. Takaki thought that the few deaths on the second ship had something to do with the protein content of the sailors' diet, but he was wrong.

About fifteen years later a Dutch physician in the East Indies, Christiaan Eijkman, noted that chickens fed mostly polished rice also contracted beriberi but recovered when fed rice polishings (i.e., the hulls). He erroneously thought that the starch in the polished rice was toxic to the nerves. Finally, Casimir Funk, a Polish chemist, showed that an extract of rice hulls prevented beriberi. He thought that this substance fell into the chemical category of "amines," and since it was "vital" to life he called it "vitamine." When the substance turned out not to be an amine, the final *e* was dropped.

A short time later, E. V. McCollum and Marguerite Davis at the University of Wisconsin discovered that rats given lard as their only source of fat failed to grow and developed eye problems. When butterfat or an ether extract of egg yolk was added to their diet, growth resumed and the eye condition was corrected. McCollum suggested that whatever was present in the ether extract be called fat-soluble "A," and that the water extract Funk had used to prevent beriberi be called water-soluble factor "B." When the water-soluble extract was found to be a mixture of compounds, its components were given designations with numerical subscripts. The specific antiberiberi factor was eventually called "vitamin B_1," or "thiamine." These "vitamins" had a common function. They formed part of the various enzyme systems needed to metabolize proteins, carbohydrates, and fats. Some of the compounds in Funk's water extract eventually turned out to offer no protection against any specific disease and their names had to be removed from the list of vitamins. As other water-soluble substances required by the body were discovered, they were added to the B-vitamin list.

Other vitamins were subsequently identified and given the designations "D" and "E" in order of their discovery. Vitamin K was so called because its discoverer, the Danish biochemist Henrik Dam, proposed the term *koagulations vitamin* because it promoted blood coagulation. Are there still unrecognized vitamins? Not likely. Patients have now been kept alive for many years through total parenteral nutrition (TPN), which involves using an intravenous formula that incorporates the known vitamins. No nutritional-deficiency diseases have shown up in spite of the fact that no vitamin B_{15} is to be found in the formula.

So what is vitamin B_{15}? Essentially a scam. In the 1950s, Dr. Ernst T. Krebs came up with the idea that a compound extracted from apricot pits, called "amygdalin," was able to selectively target cancer cells and destroy them by releasing cyanide. Krebs and his son, Ernst Jr., became the first proponents of administering this compound, an approach called "Laetrile therapy." When the government began to ask for evidence that the drug actually worked, Krebs changed his approach. The public was becoming familiar with the benefits of vitamins so he decided to convert Laetrile into one. Krebs then claimed, without any evidence, that cancer is caused by a deficiency of vitamin B_{15}. Numerous studies carried out since have failed to show that this substance can treat or prevent any type of cancer. Whatever amygdalin may be, it is not a vitamin.

And what happened in the finale of *The Blob*? The authorities eventually figured out that the menacing goo couldn't stand the cold. So the air force found a way to transport it to the Arctic and put it in a deep freeze. Which is precisely what should be done to the unsubstantiated claims being made about vitamin B_{15}.

Toxic E-mail

I bring sandwiches sealed in those special little plastic bags to work for lunch. I cook vegetables in the microwave oven in bowls covered with Saran Wrap. I microwave leftovers in Tupperware and store my cheese in cling wrap. And I do all of this fully aware of Claire Nelson's award-winning science fair project and Dr. Edward Fujimoto's comments on Hawaiian TV.

I mention my dining habits in response to a pass-this-on-to-your-friends e-mail that has circulated widely, alleging that cancer-causing substances such as plasticizers and dioxin leach out of plastics and that "Saran Wrap placed over foods as they are nuked with high heat actually drips poisonous toxins into the food." The overall message is to be wary of plastic wraps and to stay away from plastics in the microwave. Are such worries warranted? For the answer we must, as always, look at the facts.

The warning e-mail begins with the captivating saga of Claire Nelson, an inquisitive high-school student in Arkansas who learned that a plasticizer called "di(ethylhexyl)adipate," or DEHA, is found in plastic wrap and that the US Food and Drug Administration (FDA) had never studied whether this "carcinogen" migrates into food during microwave cooking. With the help of a professional scientist she devised an experiment in which she cooked up a mix of plastic wrap and olive oil and found that DEHA migrated into the oil at levels far higher than the FDA standard of 0.05 parts per billion. Claire eventually received the American Chemical Society's top prize for students, and her story charmed many a reporter. They were keen to portray her as a people's champion who had uncovered yet another attack on the public's health by an uncaring industry aided and abetted by an incompetent FDA.

Claire Nelson is a real student and she did win a prize for her work. But the prize was for her systematic investigation of a possible problem, not her role in unmasking a cancer threat. Indeed, there was no threat to be unmasked, as the migration of DEHA into food had been studied before. The notion that Miss Nelson was the first to think of this idea is romanticized folklore. After all, the FDA already had a standard in place for acceptable levels of DEHA, the standard that Nelson's results exceeded. Her finding actually came as no great surprise, considering her method. Heating plastic wrap immersed in oil for extended periods to study the migration of plasticizer is hardly a realistic situation. It is akin to trying to evaluate the risks of city driving by studying Formula 1 racing.

In any case, are plasticizers as dangerous as portrayed in the science fair project? These chemicals are commonly added to plastics to make them soft and pliable. Shower curtains, for example, contain plasticizers. Plasticizers are also used to improve "cling" in certain food wraps. Concerns have arisen over some of these chemicals because of the possibility that they may have estrogen-like properties (particularly di[ethylhexyl]phthalate [DEHP]), which, in theory, can be linked to certain cancers. But DEHA, the plasticizer used in polyvinyl chloride (PVC) wrap, does not fall into this category. Both the European Union and the US Environmental Protection Agency (EPA) have now classified DEHA as "not a suspected carcinogen." This is the plasticizer that Claire Nelson studied.

But there's more: Only PVC wraps are plasticized with DEHA. While these wraps are commonly used in commercial food packaging, they are not the ones consumers are likely to purchase and use in their microwave ovens. GLAD Wrap, for example, is made of polyethylene and contains no plasticizer. Saran Wrap uses acetyltributyl citrate as its plasticizer, a citric acid derivative that has not been implicated in any problems. So I really can't imagine what "poisonous toxin" (are there any nonpoisonous toxins?) "drips into food" from Saran Wrap. It does stand to reason that any plastic wrap should be kept out of direct contact with food in a microwave for the simple reason that food, particularly if it has high sugar or fat content, can get very hot and melt the plastic. Eating melted plastic may not be dangerous, but it will surely be unpalatable—although not as unpalatable as Dr. Fujimoto's questionable comments about microwaving foods in plastic containers.

This gentleman's words spread like wildfire when some unknown scribe immortalized them on the Web after watching an interview with Fujimoto on a Hawaiian TV station. If the reports about his comments are accurate, Dr. Fujimoto's chemical and toxicological expertise is highly suspect. He alleges that heating food in plastic containers in the microwave will cause the transfer of carcinogenic dioxins into the food. Fujimoto is right about one thing: dioxins are carcinogens, and we have to make every effort to avoid them. But he's wrong about the microwave connection. For plastics to release dioxins, two conditions have to be met: they must contain chlorine and they must be heated to incineration temperatures. The containers consumers use at home (e.g., Tupperware, GLAD Ware, Rubbermaid) are made of polyethylene or polypropylene and cannot give rise to dioxins for the simple fact that they do not contain chlorine. Nor do the containers in which you bring home those delicacies from the deli counter; these are also usually made of polypropylene. As a general rule, such containers, including old margarine tubs, should not be used in the microwave—not because of any dioxin issue, but because they may soften or melt.

The only kind of common container that could, in theory, generate dioxins is one made of PVC. While PVC is used extensively in cleaning products and cosmetics packaging, it is not used to make microwavable food containers. Even if it were, temperatures in the microwave are not nearly high enough to break the plastic down and form dioxin.

Fujimoto underlined the dioxin risk by recalling how fast-food restaurants switched from foam containers to paper wrappings when they learned about the connection between plastics and dioxins. Nonsense. The foam containers were made of polystyrene, so there was no dioxin risk. Environmental concerns about freons used in the manufacture of polystyrene foam prompted the switch.

Basically, then, in spite of the alarming e-mail that may taint your in-box, there is no scientific basis for concerns about using plastics in the microwave. There is reason, however, to be concerned about the ease with which unreliable information spreads via the Internet and the unnecessary anxiety it creates. I recently had a panicky caller ask me if it is true that "shower curtains can cause cancer." He had heard "that some student had discovered this." Can you guess which student's science fair project germinated this rumor?

NATURAL OR SYNTHETIC?

Now, repeat after me: "The properties of a substance depend on molecular structure, not ancestry. When it comes to assessing effectiveness and safety, whether the substance is synthetic or natural is totally irrelevant." I bring this issue up because the belief that "natural is better" is so widespread and so . . . wrong! One–one thousandth of a milligram of botulin toxin will kill an adult. And it's a perfectly natural substance! Scorpion venom, cocaine, ricin, nicotine, morphine, and a myriad of other naturally occurring substances are remarkably toxic as well. They are not toxic because they are natural; they are toxic because the molecules of which they are composed just happen to interfere with some aspects of body chemistry. Of course, such interference is not necessarily negative. Although morphine is highly toxic, at appropriate doses it is a wonderful painkiller. And so it goes for synthetic substances. The safety and efficacy of fluoxetine (Prozac), ibuprofen (Advil), or ascorbic acid (vitamin C) are functions of the molecule's chemical structure, not its origin.

Surprised by the mention of vitamin C as a synthetic substance? Well, almost all of the vitamin C sold in the world is synthetic. In the most common process, glucose is treated with hydrogen at high pressure to convert it to sorbitol, which is then fermented with the bacteria *Acetobacter xylinum* to yield sorbose. A series of chemical reactions then converts sorbose to vitamin C. This vitamin C is the same in every way to the one found in fruits and vegetables. It is "nature-identical." While isolation of vitamin C from natural sources is certainly technically possible, it is financially prohibitive and functionally unnecessary. Simply stated, what matters is the final molecular structure, not how that structure is achieved. Some drugs are isolated from natural sources and used unaltered while some are isolated and chemically modified and others still are synthesized from scratch. Modern pharmacology is a fascinating blend of natural and synthetic. The "statin drugs" used to reduce blood cholesterol levels are a good case in point.

While many people think of cholesterol as the villain that causes heart disease, the fact is that there would be no life without it. Cholesterol is an important component of cell membranes and is the precursor for important biomolecules such as testosterone and bile acids. There is, however, no requirement for cholesterol in the diet since our bodies are quite capable of synthesizing it. As early as 1910 researchers recognized that cholesterol also had a nasty side: it is present in the arteriosclerotic plaques that clog arteries and cause heart disease. Understandably, then, scientists became interested in the mechanism by which cholesterol was made in the body, hoping to explore possible ways to limit its synthesis and perhaps reduce the risk of heart disease. Over a period of some twenty years, beginning in the late 1940s, researchers learned that most cholesterol synthesis takes place in the liver, where in a sequence of over twenty steps, a series of enzymes pieces the molecule together. The critical step in the synthesis involves an enzyme with the tongue-twisting name of "hydroxymethylglutaryl coenzyme A reductase" (HMG CoA reductase). If the activity of this enzyme could somehow be impaired, cholesterol synthesis could be reduced, perhaps offering some benefit to people with high blood levels of cholesterol. But where would one search for a substance that could interfere with a cholesterol-synthesizing enzyme?

Researchers at Sankyo Pharmaceuticals in Japan had an idea. They knew that many plants and fungi produce toxins to deter animals from eating them. Poison ivy and poisonous mushrooms are legendary examples. Akira Ando and colleagues reasoned that since herbivores rely on cholesterol biosynthesis, plants or fungi might have developed protective mechanisms by producing chemicals that interfered with such synthesis. These plants or fungi would then be toxic to noncarnivorous animals. So the researchers began a systematic investigation of organisms that could conceivably produce HMG CoA reductase inhibitors. Molds of the *Penicillium* species were good candidates because they were, of course, already known to produce compounds that interfere with life (i.e., "antibiotics"). The idea turned out to be a good one, and in 1976 Ando isolated a compound from a mold fermentation broth that turned out to be an effective inhibitor of cholesterol biosynthesis. "Compactin," as the compound came to be called, offered great hope for reducing blood cholesterol levels. Unfortunately, however, due to numerous side effects, the hope was not realized. But compactin did clearly demonstrate the principle of reducing cholesterol by means of HMG CoA reductase inhibitors. The challenge was to find a substance with a better safety profile.

Numerous pharmaceutical companies with visions of large profits dancing in their heads quickly mounted research programs to find an improved compactin. Merck hit pay dirt first with lovastatin (Mevacor), another fungal metabolite. Then, as is usually the case, chemists went to work altering the molecule in the laboratory, hoping to come up with an even better version. Such efforts yielded simvastatin (Zocor), which can be appropriately labeled a "semisynthetic" compound. Of course, many compounds are synthesized before a successful one is found, and chemists slowly learn the structural features a molecule needs for efficacy and the ones responsible for side effects. Then comes the ultimate challenge of making a molecule from scratch that incorporates the desirable features. One of the most popular statins, Pfizer's atorvastatin (Lipitor), is an example of a totally synthetic statin that has an excellent risk-benefit profile. On the other hand, Bayer's cerivastatin (Baycol), which had to be taken off the market because of side effects, was also a synthetic statin. Compactin, the prototype statin, was never marketed because of unacceptable side effects. And it was "all natural." So repeat after me again: "The properties of a substance depend on molecular structure, not ancestry. . . ."

Count Your Way to Good Health

Would you like to reduce your risk of cancer? You can, if you can count to five. If you can make it to ten, you'll reduce your risk even more. The secret is not in the mental gymnastics; it's in what you will be counting. I am constantly amazed by people who swallow a surfeit of unproven dietary supplements, hoping to protect themselves from cancer. Instead, they could be following a relatively simple regimen that has been scientifically shown to reduce risk by up to 50 percent. I am talking, of course, about consuming five to ten servings of fruits and vegetables every day.

Scientists, as you know, waffle on many issues. We're not sure about the risks associated with radiation from cell phones, mercury in fish, or fluoride in water. We debate the extent to which greenhouse gases need to be controlled. We puzzle over the use of estrogen supplements and often don't agree on the pros and cons of genetically modified foods or pesticides. This vacillation may seem surprising, because the scientific literature on each of these issues is abundant. But these are complex topics, with enough "on the one hand" and "on the other hand" arguments to preclude categorical conclusions. It is rare indeed to find a subject on which there is virtually universal scientific agreement.

Rare, but not impossible! One would be hard-pressed to find a scientist who opposes increased consumption of fruits and vegetables. Why? Because the evidence is as ironclad as is possible in the often-confusing world of human nutrition.

What constitutes such strong evidence? Hundreds and hundreds of studies carried out around the world have furnished us with data demonstrating that fruit and vegetable intake reduces the risk of cancer. An analysis of these studies serves to prove the point and also allows us to learn about the various types of studies that scientists design in order to be able to offer meaningful nutritional advice.

Information about a potential link between diet and disease usually first emerges from what is called a "cohort study." Detailed food frequency and health status questionnaires are mailed to thousands of healthy people periodically over many years; the responses are then analyzed to see if nutritional patterns can be linked with specific diseases. A classic example is the Nurses' Health Study, which began in 1976 when 121,700 female nurses aged thirty to thirty-five started to fill out biennial questionnaires. After just ten years it became apparent that subjects who consumed the most carotenoid-containing fruits and vegetables had a 20 to 25 percent lower risk of lung cancer than those who ate smaller amounts. Carotenoids are familiar as the yellow and red pigments found in carrots and tomatoes. In another cohort study, over 1,200 Massachusetts residents over the age of sixty-five were asked to report on their diets in 1985 and were then followed by researchers. Those who eventually died from cancer ate the fewest vegetables and those who ate the most green and yellow vegetables had the lowest cancer rates.

The most plentiful information about the link between diet and cancer comes from "case-control studies." Persons with a particular disease are identified and asked about their past dietary history and other possible risk factors. They are then matched against a similar group of "control subjects" who do not have the disease. In a typical case-control study, 179 pancreatic cancer cases were matched with 239 controls. Patients with the disease were more likely to have consumed more smoked and fried foods and far less raw fruits and vegetables. Over 150 such studies have shown that fruits and vegetables provide a protective effect against various cancers. The greatest evidence of this effect is for cancers of the stomach (the most common cancer in the world), lung, and esophagus.

Animal-feeding studies and test-tube experiments also offer insight. The animals in question are usually rats or mice that are fed a diet of certain fruits or vegetables before they are exposed to a chemical carcinogen. Their tendency to develop tumors is then investigated. Perhaps the most alluring example here is one we have already encountered (see "For the Love of Broccoli" on page 107). Dr. Paul Talalay, director of the Brassica Chemoprotection Laboratory at Johns Hopkins University in Baltimore, Maryland, was stimulated by a number of cohort studies that showed an inverse correlation between cancer and the consumption of "cruciferous" vegetables, such as cauliflower and broccoli. He isolated a compound from broccoli that appeared to have anticancer properties. He then fed this newly isolated compound, called "sulforaphane," to rats that had been exposed to dimethylbenzanthracene, a potent carcinogen found in smoke. Almost 70 percent of the control rats developed cancer, but only 35 percent of the animals that had dined on the broccoli extract did. Dr. Talalay then added sulforaphane to cultured human cells in the laboratory and again demonstrated protection against cancer. He was so taken by his results that he began a systematic search for the best dietary source of sulforaphane and found it in a species of broccoli sprouts that he patented as BroccoSprouts. Each gram of these sprouts is guaranteed to have as much sulforaphane as 20 grams of mature broccoli. A preliminary estimate is that just a handful of these sprouts per week can cut the risk of colon cancer in half.

We are even beginning to see some human "intervention studies" using vegetables. These investigations are potentially the most compelling. Because of the reputed protective effect of tomato products against prostate cancer, fifteen patients at Wayne State University in Michigan were given 15 milligrams of lycopene, the red pigment in tomatoes, twice daily for three weeks prior to surgery. The patients' tumors were smaller and showed reduced malignancy compared with the tumors of patients who had been given a placebo.

While no single study can, or should, be convincing about the protective effect of fruits and vegetables, there is no doubt that the preponderance of evidence is overwhelming. Still, most North Americans consume three or fewer servings per day. If they would only double their intake of fruits and vegetables instead of searching for solace in the latest multilevel-marketed "miracle," we would likely see a drop in cancer incidence. So all together now, let's count: (1) a glass of orange juice; (2) a carrot; (3) half a cup of broccoli; (4) half a cup of grapes; (5) a banana; (6) an apple. That's not so hard to do, is it? And don't forget the tomato sauce. You can add some garlic to it for further protection against cancer—and vampires.

SOME NEAT CHEMISTRY

Wondrous Tricks and Bad Smells

The curtain goes up in Heller's Wonder Theatre on Broadway in New York to reveal a blindfolded woman seated comfortably center stage. A volunteer from the audience is solicited to "pick a card, any card!" She does so, thinking, of course, that it is a free choice. It isn't. Robert Heller, the premier American magician of the mid-nineteenth century was highly skilled at "forcing" a card. There was nothing really novel about that skill, but his revelation of the selected card was truly innovative. Heller resorted to chemical magic! But back to his show. He silently approaches the blindfolded, bare-armed woman and waves his hands above her. Suddenly, a blood-red image appears on her arm, an image of the card that had been selected! The audience bursts into thunderous applause.

I learned about Heller's chemical conjuring from a marvelous book I received as a gift. *Scientific Mysteries and How to Produce the Most Interesting Chemical, Optical and Physical Illusions* is for me a true treasure. Published in 1891 in London, England, the little volume is a compilation of various scientific effects used on the stage to titillate Victorian audiences. Not only are some of the "tricks" truly ingenious, but they also show

clearly that those of us involved in performing chemical magic today may just be reinventing the wheel—a sobering thought.

Recently I purchased a trick knife designed to create the illusion that it slices halfway through an arm. It came complete with a hidden bulb that can be filled with red food dye, which squirts out at the appropriate moment. You may be wondering why I was interested in such a gross item. There was some method to my madness. I had in mind to incorporate this illusion into a lecture on the history of charlatans, having learned that at one time mountebanks performed the trick and then pretended to heal their wounds with whatever wondrous nostrum they were selling. Actually, I'm not sure how they produced the "blood," because the gimmick really didn't work well. The red dye didn't seem to come from the wound, and worse than that, it made a big mess. So I put on my thinking cap and worked on finding a chemical solution to the problem. The challenge was to create the illusion of drawing blood with a knife.

Chemical analysis often involves various color tests. Starch, for example, reacts with iodine to produce a deep blue color; chlorine with toluidine produces yellow; and iron with potassium thiocyanate (KSCN) produces a "blood" red color. The latter is a very sensitive reaction and is commonly used by criminologists to test for footprints. Most soils will contain some iron compounds, which stick to the bottoms of shoes. Footprints that are invisible to the naked eye can therefore often be visualized when sprayed with a solution of potassium thiocyanate. Thinking that I was very clever, I decided to apply this reaction to my knife problem. I made a dilute solution of iron chloride ($FeCl_3$) and rubbed some on my arm. After it dried, it became invisible. Then I dipped the knife in a solution of potassium thiocyanate and was ready for the effect. I even had some patter for this maneuver, suggesting to the audience that the knife had to be disinfected before I proceeded with the dangerous experiment. Lo and behold, it worked! The illusion was amazing. The knife seemed to slice right into my arm, producing "blood" appropriately. I was proud of my ingenuity.

Now, of course, I see that it had been thought of before. My recently acquired 100-year-old book describes Robert Heller's illusion in detail. Using a camel-hair brush dipped in "tincture of perchloride of iron," he drew the desired image on his stage assistant's bare arm. In his hand the magician hid a small bulb filled with a solution of potassium sulphocyanide (as potassium thiocyanate was called at the time), which he sprayed on the arm while waving his hands around in the time-honored fashion of illusionists. And why did the woman have to be blindfolded? Simple. To protect her eyes from the chemical spray!

As it turns out, this particular chemical trick was not the only one performed by magicians of the era. My *Scientific Mysteries* describes a neat little illusion known as the "Wonderful Bottle." An illusionist would tilt a brown glass bottle to show it was empty, then ask an audience volunteer to fetch some water to fill it. Then, in response to requests from the crowd, the magician would pour out port wine, claret, milk, champagne, or ink. Everyone would be flabbergasted! Actually, the bottle used in this trick wasn't quite empty; it contained a small amount of ferric chloride dissolved in hydrochloric acid. When the solution formed by adding water to this mixture was poured into a glass that had a few drops of a concentrated potassium thiocyanate solution in it, "port wine" resulted. A dilute solution of thiocyanate yielded "claret." Lead acetate reacts with ferric chloride to produce insoluble white lead chloride. So when the ferric chloride solution was added to a glass with a bit of lead acetate at the bottom, a white precipitate that passed for milk was produced. Magicians created "champagne" when hydrochloric acid reacted with baking soda (sodium bicarbonate) at the bottom of the glass. Treating baking soda with an acid produces copious amounts of carbon dioxide gas, the bubbles in the "champagne."

And the ink? That was easy as well. Ferric sulfide is black, so pouring the ferric chloride solution into a glass treated with ammonium sulfide resulted in "ink." It undoubtedly also resulted in a rather disturbing smell. Ammonium sulfide readily releases hydrogen sulfide, which is the memorable odor of rotten eggs. It is also a nose-wrinkling component of human flatus. This notion has not been lost on the manufacturers of some modern novelties. "Fart bombs" are packaged in little foil pouches that contain bicarbonate, ammonium sulfide, and a bag of vinegar. Squeezing the pouch releases the vinegar, which combines with the bicarbonate to generate carbon dioxide gas. In about ten seconds enough pressure builds up to explode the foil pouch and release the fragrance of hydrogen sulfide. A novel discovery? Hardly. Leonardo da Vinci designed stink bombs that could be launched with arrows! (I'll bet you didn't know he had chemical inclinations!) *Plus ça change, plus c'est pareil.*

SEX AND SCENTS

When Leopold Ruzicka received the Nobel Prize for chemistry in 1939, African civet cats and Asian musk deer collectively breathed a sigh of relief. Professor Ruzicka was honored for having determined the molecular structure of muscone and civetone, the prized perfumery ingredients that had cost numerous such animals their lives. With the structures of these molecules determined, they, along with various analogs, could probably be synthesized in the laboratory.

Musk is the most valuable animal product that exists. It is far more valuable than gold. The male musk deer, a shy animal usually weighing no more than 25 pounds, features a small hairy pouch on its abdomen. In it is the highly valued waxy, reddish substance that the animal discharges along with its urine during rutting season to mark its territory and attract females. But this "musk" does something else as well: it adds a "sensual" note to perfumes and acts as a "fixative," making the scent of other ingredients linger longer. This property made musk an important item of commerce by the seventh century, and made the capture of musk deer financially lucrative. Unfortunately, the traditional method for harvesting musk was to trap the animals and kill them before cutting out the valuable glands.

The smell of concentrated musk is actually putrid, but at great dilutions it becomes pleasant, especially as far as women are concerned. Their sensitivity to musk varies with their menstrual cycles, peaking at the time of ovulation. It is noteworthy that male sweat contains some compounds that have a musky fragrance, so perhaps natural musk just happens to resemble chemicals that men produce and women find attractive. The main value of musk, however, remains its ability to act as a fixative in perfumes. Robert Boyle, one of the leading scientific lights of the seventeenth century, recorded his amazement that a pair of scented gloves he had possessed for some thirty years still smelled! The secret to the scent's staying power was musk. Or, more specifically, muscone; as we learned in 1906, it was the active ingredient in the male musk deer secretion. This was the compound that was the focus of Professor Ruzicka's research. But it wasn't the only such substance he was interested in. Ruzicka also knew that the African civet cat produced a similar-smelling secretion, one that also had a long history in perfume manufacture, and he wondered if this active ingredient was similar to that found in the musk deer scent pods. Civetone indeed turned out to be similar to muscone; both compounds consist of carbon atoms joined together to form large rings.

The history of civetone is as disturbing as that of muscone. In this instance, though, the animals aren't killed. Instead, they are kept in cages so small that they can't turn around. The cages are equipped with doors at the rear that afford access to the animals' anal scent glands, which are scraped out every few days with a special spoon to yield the prized perfume ingredient. Needless to say, most countries are disturbed by the killing of musk deer and the tormenting of civet cats and do not trade in natural muscone or civetone. France and Japan, however, still import musk, mostly from China, where musk deer are raised on farms. There is also a huge black-market trade in musk and civet gland extracts, which may be understandable given the state of certain local economies. A few musk pods can fetch enough money to allow a family to live for months. The problem is exacerbated by the Chinese belief that musk is a cure for a variety of ailments. There is absolutely no scientific evidence for this claim, but the poaching of musk deer for the supposed medicinal value of the glands continues.

Had chemists not stepped into the picture in the early 1900s, civet cats and musk deer may be extinct today. Ruzicka's determination of the structure of muscone and civetone allowed for the synthesis of these compounds in the laboratory. But the syntheses were difficult and not commercially viable. Then a lucky accident occurred. Wallace Carothers, the brilliant DuPont chemist who would eventually discover nylon, was working on building some giant molecules called "polyesters." He noted that the reaction produced some side products that had interesting fragrances. Carothers asked his assistant, Julian Hill, to look into the matter. Hill did so, with increased vigor after he came home one day and aroused his wife's suspicions because he smelled of perfume. Indeed, the side product of the polyester reaction turned out to be a ring compound similar to muscone. DuPont jumped on this happy finding and began to market Astrotone, the world's first synthetic musk.

Today there are numerous such substances. Researchers have determined the structural features a molecule must have to smell musky, and hundreds of candidates have been produced. Many have made it into commercial products. In fact, they have become truly ubiquitous. The presence of musk is no longer limited to perfumes; laundry detergents, cleaning agents, and even some foods contain musks. Whenever the intensity and duration of a fragrance needs to be amplified, there is a synthetic musk ready to do the job. Perhaps too ready. Many of these musks degrade with difficulty and become persistent in the environment. Studies have shown that synthetic musks can be found in waterways and accumulate in fish. Nobody knows if this persistence has any consequences for fish or humans. Some people allege that musks have estrogenic effects or cause cell mutations or respiratory problems. Others maintain that musk used as aromatherapy enhances health and acts as an aphrodisiac. Take your pick—there is no significant scientific evidence for any of these claims.

While synthetic musks abound, there is still demand for perfumes made with the natural material and rumors persist that some of the most expensive perfumes are made with natural muscone and civetone. Producers say they use synthetic versions of these compounds. Are they telling the truth? Actually, there are subtle differences between the natural and synthetic compounds. Synthetic muscone exists in two forms, with the molecules being nonidentical mirror images of each other. Chemists can readily determine this nuance and modern chemistry may once more come to the rescue of the musk deer and the civet cat.

From "Yuck" to "Yum"

Soon you may be seeing a new name on the ingredients list of many processed foods. You may also see it on the label of a bottle on the spice rack in you supermarket. The name of this new ingredient? Adenosine monophosphate. Oh, I can already see the bitter expression forming on many a face out there. I know what many readers are thinking: "There goes big business again, tainting our food supply with yet another additive just for the sake of profits." Not so. First, you should understand that introducing a new food additive is not an easy business; numerous safety studies are needed for market approval. Second, you should know that food additives perform important functions. In this case, adenosine monophosphate (AMP) may be just the substance to wipe that bitter expression off your face. Why? Because this chemical can reduce the bitter flavor of many foods. And that, in turn, can reduce the risk of cancer! How? Let's start at the beginning.

In this case the beginning is 1931 in a research lab at the DuPont chemical company. Arthur Fox was working on a developing a new artificial sweetener when one of his chemicals, phenylthiocarbamide (PTC), accidentally scattered into the air. Fox didn't think much of it, but one of his colleagues sure did. He inhaled a little of the chemical and found it to be one of the most bitter flavors he had ever encountered. This experience aroused Fox's curiosity, and he tasted some of the stuff himself—with absolutely no result. None! To him, PTC had no flavor at all. What was going on here?

Fox enlisted other colleagues to sample PTC and it became clear that at least as far as this chemical went, the world was divided into nontasters and supertasters. At the slightest exposure to PTC, the latter scrunched up their faces in horrific grimaces. It turns out that the ability to taste PTC is inherited, with about 25 percent of the population falling into the supertaster category. So what? you may ask. Outside of a chemical catalogue, PTC is not commonly encountered. True. But most people are likely to encounter broccoli, Brussels sprouts, soy products, and green tea. All these foods contain naturally occurring chemicals that in supertasters can elicit the same kind of revulsion as PTC does. This time, though, there is a consequence: the bitter flavor supertasters experience may encourage them to stay away from these foods in spite of established health benefits.

Linda Bartoshuk of Yale University in New Haven, Connecticut, is probably the world's leading expert on supertasters. She was the first to show that their tongues have special anatomical features. Supertasters have more fungiform papillae, the tiny mushroom-shaped structures that house the taste buds. Basically this means that supertasters have more taste buds than regular tasters. But the story gets even more interesting. Bartoshuk decided to investigate whether dietary choices made by supertasters had any effect on disease patterns. In a preliminary study of colon cancer patients, she learned that they were more likely to be supertasters. In fact, the number of cancerous polyps they had was related to the extent that they perceived certain foods as bitter. She also found that supertasters were more at risk for gynecological cancers, again because they probably avoided certain protective foods, such as soy. On the other hand, supertasters also find the taste of sweets and fats more intense and tend to eat less of these, putting them at reduced risk for cardiovascular disease. But how could they be encouraged to eat more of the foods they perceived as bitter?

Enter Linguagen, an American biotechnology company that has come up with a family of compounds that are best referred to as "bitter blockers." When mixed with foods, these compounds can neutralize unpleasant tastes. The discovery hinged on the observation that molecules responsible for bitter tastes fit into receptors on the taste buds and trigger the release of a protein called "gustducin." This release sets in motion the biochemical machinery that eventually results in nerve impulses being sent to the brain to signal the perception of bitterness. Linguagen decided to investigate the possibility of finding molecules that would block the bitter receptors on taste buds. Not only were test-tube experiments successful, but Linguagen researchers also learned, to their great delight, that the candidate molecules occurred naturally in various foods. Approval for a food additive would therefore be greatly facilitated. The molecules that blocked the release of gustducin turned out to be nucleotides, compounds that are the building blocks of DNA. Specifically, adenosine monophosphate (AMP) seemed to hold the greatest potential.

Test-tube experiments were one thing, but would AMP really alter the taste of foods? As is often the case in such matters, Linguagen scientists turned to rats for guidance. As a rule, rats will avoid water laced with caffeine because of the compound's bitter taste. But when AMP was added to the brew, the rats did not distinguish between caffeinated and regular water. With this success it was time for human trials. And in this case, the rats turned out to be a good model for human behavior. The classic bitter taste of naringin, which is found in grapefruit juice, was cut when the juice was treated with AMP. This was a very useful observation given that naringin is believed to fight cancer. The next step will be determining if the taste of broccoli and other vegetables that normally elicit a "yuck" can be improved with a sprinkling of AMP. Such an improvement would certainly go a long way toward increasing the proportion of the population that consumes the five or more servings of fruits and vegetables per day, which many researchers think can halve the risk of certain cancers.

There are still more benefits that "bitter blockers" may provide. Many medications have a very bitter taste that discourages compliance, especially in children. We all know that a spoonful of sugar makes the medicine go down, but this practice is not always an option. In theory at least, AMP can be incorporated into many medicines. It could also be used to reduce the vast amounts of sugar, salt, and fats that are added to processed foods. In many cases these substances are used to mask the bitterness associated with some of the naturally occurring compounds in food or those created by processing. If AMP had been around during George Bush Senior's presidency, he might not have had to declare his aversion to broccoli and could have avoided the ugly scene of angry farmers dumping piles of broccoli in front of the White House. Now, do I see some of those bitter expressions generated by the prospect of a new food additive changing into smiles?

Nibbling on Glass

Occasionally I like to nibble on glass for dessert. I don't mean that I chew on broken bottles. The glass I fancy adorns the famed Hungarian Dobos Torte, and it's made of sugar! This fact may come as a bit of a surprise, as most people assume glass is made by heating sand. But that is not always the case. The term *glass* doesn't describe a specific substance; it describes a state of matter. If you want to get technical about it—and why not—then glass can be described as a substance cooled below its melting point without crystallizing.

To understand what this definition means, let's first look at water. As the temperature drops to 0°C, water freezes into ice. On the molecular level, the H_2O molecules align themselves into a rigid three-dimensional pattern characteristic of crystalline materials. The pattern is organized and repeats in every direction. When ice melts into water, this pattern is disrupted and the molecules begin to move about randomly. In the case of solid sugar, the molecules stack together neatly, just like they do in ice. And just like ice, sugar has a distinct melting point. But if melted sugar is cooled quickly, the molecules do not have a chance to organize themselves into a crystalline pattern. This is because unlike water molecules, sugar molecules are large and nonsymmetrical. Basically, they become frozen in place in the disorganized pattern that characterized their liquid form. Water can never form a glass because its small, symmetrical molecules are eager to form organized, repeating units.

In glass, then, the molecules are arranged as they are in a fluid, with no long-range order, even though the glass is solid. It will soften with heat and eventually melt, but not at a specific temperature. Candies such as Life Savers are made of this type of glass, and so is the candy coating on the enticing slice of Dobos that sits in front of me as I write.

Making candy glass takes experience; if the cooling isn't done correctly, the candy will crystallize and the "glassy" effect is lost. If you really want to do this right, you can seek help from a prop designer. Remember all those films that had cowboys diving through glass windows and smashing bottles over each other's heads? Those stunts were all done with candy glass. Pour melted sugar into an appropriate cold mold and—presto! You have bottles that won't hurt even the most soft-headed actors. The catch is that to learn the technique, you'll have to find an old-time prop man. Today, most fake glass bottles are made of specialty plastics. Indeed, many plastics form glasses. Consider polystyrene, for example. That's the stuff used to make those clear plastic "glasses" without which air travel would not be possible. Polystyrene is composed of very long molecules that readily tangle, making liquid polystyrene something of a viscous soup. In general, the more viscous a liquid is at its freezing point, the more trouble the molecules will have forming crystals.

Let's move on to real glass. The type made from sand. In all likelihood, volcanoes produced the first glass samples humans ever saw. Obsidian was a dark glass that formed when silicate minerals in lava cooled and solidified. The first effective cutting tools were probably made of obsidian. Knowing where this substance was found, it would have been quite natural for people to try to simulate nature and attempt to heat various minerals and then cool them to produce glass. They likely tried all sorts of substances and eventually discovered that sand was the best candidate. Pliny, the Roman historian, recorded his version of the story in the first century A.D., but given he speaks of events that happened some 4,000 years earlier, his account may not be reliable. In any case, Pliny maintains that the Phoenicians were the first people to produce glass when they built fires on a beach and supported their pots on chunks of natron, or sodium carbonate. The sand and natron fused under the high heat and yielded glass upon cooling. This story is probably apocryphal, but we do know that by about 1500 B.C. the Egyptians were making glass bottles. Adding limestone, or calcium carbonate, lowered the temperature needed to melt the mix. And that basically is still the process used today. Sand, soda, and limestone are melted together and the mix is cooled to form glass.

Of course, that's an oversimplification. We have various specialty glasses that require specific modifications to the basic technique. Tempered glass, for example, is extremely strong and is made by reheating the glass to about 600°C and then quickly quenching it with high-pressure cold air. This process almost instantly contracts the surface, packing the molecules together tightly and increasing the strength dramatically. Since the outside of the glass contracts so much more quickly than the inside, a certain amount of inner tension is generated. If the glass breaks, this tension is released and the glass shatters into tiny pieces. Accordingly, tempered glass is used on the side windows of cars, in glass doors, and for some decorative ware. When these items shatter, no sharp pieces are produced. On rare occasions tempered glass can shatter spontaneously. Since the inside is under tension, a slight change in temperature can lead to just enough of an increase in stress to cause a mighty bang. No ghostly explanations need to be invoked.

The newest development in glass technology aims to produce self-cleaning windows. In a proprietary process, a thin layer of titanium dioxide is deposited on the glass. This process does two things: First, titanium dioxide acts as a catalyst, allowing ultraviolet light to convert oxygen to an extremely reactive form known as "singlet oxygen." This singlet oxygen reacts with grime and grease to form water and carbon dioxide. In addition, titanium dioxide prevents water from beading. Consequently, rain that hits the window forms sheets of water and washes the dirt away. If all goes well, we won't have to worry about washing windows in the future!

Speaking of worry, you've probably spent a few sleepless nights pondering whether glass, which has some commonality with fluids, flows. Yes it does! But it would take millions of years to note any difference in thickness. Windows in old cathedrals are thicker at the bottom for the simple reason that glass production was not very sophisticated and panes did not have even thickness. The thick part was placed at the bottom for structural reasons. So relax—your windows are not flowing out of their frames. But the chocolate *is* starting to flow out of my Dobos. Gotta go and eat some glass.

BAD SCIENCE, BAD JUSTICE

The emergency room physician thought he recognized the symptoms. The three-month-old boy was vomiting, breathing with difficulty, and showing virtually no reflexes. A blood test quickly revealed his blood pH to be 7.0, considerably more acidic than the normal pH of 7.4. The doctor had seen cases like this before; the symptoms smacked of ethylene glycol poisoning. Could the child have swallowed some antifreeze? A blood sample was sent to a laboratory and sure enough, it turned up positive for ethylene glycol, the main ingredient in commercial antifreeze.

The doctor ordered that intravenous sodium bicarbonate be quickly administered to neutralize the excess acid, and the boy soon recovered. But something seemed strange to the emergency room staff. The mother could not account for the presence of the ethylene glycol in her baby's blood. Furthermore, there was a history of social problems in the family. Did she try to poison her child deliberately? Was this a case of Munchausen Syndrome by Proxy, a condition in which a psychologically disturbed parent deliberately harms a child to gain attention? The hospital called in social workers, and they decided that there was sufficient evidence to place the baby with foster parents. The mother was, however, allowed visitation rights.

Eight weeks later, following a visit with his mother during which she carried out an unsupervised feeding, the young boy again began to show the same signs of illness he had previously. By the time the foster parents brought him to emergency, his blood had become so acidic that despite all efforts he could not be saved. Once again, tests by two laboratories confirmed the presence of ethylene glycol. This time, Patricia Stallings of St. Louis, Missouri, was arrested for the murder of her son.

Mrs. Stallings was pregnant when she was arrested and gave

birth in prison to a baby boy who was, of course, immediately placed with foster parents. Within a short time, the second son began to exhibit symptoms similar to the ones that had afflicted his brother. Since the mother could not possibly have poisoned this baby, a more thorough medical evaluation was carried out. This time the diagnosis was different. Instead of ethylene glycol poisoning, tests determined that the child suffered from the rare genetic disease methylmalonic acidemia (MMA).

This genetic disease occurs in roughly 1 in 48,000 newborns. Victims lack a properly functioning enzyme that is required to metabolize valine, isoleucine, threonine, and methionine, all amino acids that commonly occur in the diet. Vitamin B_{12} is a required cofactor for this enzyme, but in the so-called vitamin B_{12}–responsive forms of MMA there is for some reason insufficient cofactor available to the enzyme. As a result, methylmalonic acid, a breakdown product of the abovementioned amino acids, does not get converted to the next product in the normal metabolic sequence. It is the buildup of methylmalonic acid that acidifies the blood and causes the symptoms of MMA to appear.

Interestingly, these symptoms are very similar to those seen for ethylene glycol poisoning. Ethylene glycol is metabolized to oxalic acid in the body and thereby causes acidosis in the same fashion as methylmalonic acid. Unfortunately, the diagnosis of methylmalonic acidemia in her second child was not allowed as evidence in Mrs. Stallings's murder trial because the judge ruled that her defense attorney had been unable to come up with an expert witness to testify that the genetic disease could be confused with finding ethylene glycol in the blood. The baby, like his brother, may indeed have suffered from the disease, but in the judge's view that did not preclude poisoning. In 1991 Patricia Stallings was sentenced to life imprisonment.

The story was so bizarre that it made TV's *Unsolved Mysteries*. Drs. James Shoemaker and William Sly of St. Louis University, both researchers in metabolic diseases, caught the program and thought it too great a coincidence that Mrs. Stallings would have poisoned her first son with antifreeze and then given birth to a second son who had an illness that mimicked antifreeze poisoning. They contacted the authorities and asked that they be allowed to examine blood samples from Ryan, the boy who had supposedly been murdered.

A thorough analysis of these samples revealed that a staggering mistake had been made. Both laboratories that had carried out the original analysis had erred! Using the technique of gas chromatography, which separates the components of a mixture and displays the results as a series of peaks on graph paper, they had mistaken propionic acid, a substance found in the blood of people afflicted with methylmalonic acidemia, for ethylene glycol, the antifreeze ingredient. Testing with a superior technique that combines gas chromatography with mass spectrometry (a procedure that allows exact identification of the components of a mixture) showed the presence of propionic acid. A subsequent analysis of Ryan's blood by Yale University's Dr. Piero Rinaldo confirmed the presence of methylmalonic acid. Mrs. Stallings was released from prison and reunited with her surviving child. She promptly sued the hospital and the labs. The talk is that the out-of-court settlement ran into the millions of dollars. Justice triumphed thanks to chemistry and a little luck. Why luck? First, Mrs. Stallings was lucky to have been pregnant at the time of her arrest. Second, the odds of having a second child with methylmalonic acidemia are only one in four. And finally, an expert on genetic screening happened to be watching TV at just the right time.

Much has been learned about MMA since the Stallings affair. Recently, David Rosenblatt and Thomas Hudson, geneticists par excellence at McGill University, collaborated with Melissa Dobson and Roy Gravel of the University of Calgary to identify two genes that are linked to the vitamin B_{12}–responsive forms of the disease. This work will allow for early prenatal diagnosis of MMA so treatment—vitamin B_{12} supplementation and a low-protein diet—can be initiated. The prosecuting attorney in the Stallings case learned something too: scientific evidence is not always what it seems to be. Based on the reports of "experts," he had told the jury that speculating that Ryan Stallings died from natural causes was tantamount to speculating that some little man from Mars came to Earth and shot him full of mysterious bacteria. Well, there were no mysterious bacteria involved, but there were some mysterious genes. To his credit, after arranging for Mrs. Stallings's release, prosecutor George B. McElroy summed up well: "It is difficult at any time to stand up and say a mistake has been made, but my gosh, there's a time when it just has to be done." Right on.

Santa's Mushrooms

Where is Dasher dashing? Why is Prancer prancing and Dancer dancing? Can mistletoe do more than prompt a kiss? These are pretty interesting questions, I'm sure you'll agree.

Mistletoe has had a certain mystique about it since ancient times, probably on account of the curious way it grows. The plant is a hemiparasite, meaning it can either grow in soil or, more commonly, spring from the branch of a tree. At one time ladies probably stood under the branch in awe, admiring the pretty flowers, which gave gentlemen an opportunity to take a little liberty with the fair sex.

The original mistletoe, *Viscum album* (which is different from the ornamental North American version), got its name from the Anglo-Saxon words *mistel,* for "dung," and *tan,* for "twig." Dung-on-a-twig is an excellent description of the plant's

origin. Mistletoe would often appear on branches where birds left droppings containing mistletoe-berry seeds that had passed through their digestive tracts. Interestingly, birds are not bothered by the seeds, which are highly toxic to humans. The main culprits are viscotoxins, small proteins than can destroy cells.

Any substance that has such an effect on human health arouses scientific curiosity. Pharmaceutical history is peppered with attempts to use small doses of poisons to wipe out a disease without wiping out the patient. Arsenic, mercury, strychnine, and belladonna are obvious examples. So it should come as no surprise that various mistletoe preparations have also appeared in drug compendia. Until the 1920s, the scientific community dismissed these remedies as mere placebos. But then researchers discovered that mistletoe also harbors some complex compounds called "lectins," which can bind to cells and induce biochemical changes. Attention focused on the possibility that these substances, at the right concentrations, might selectively destroy cancer cells. Early on there was encouragement from laboratory studies and animal trials that showed a slowing of the growth of certain tumors in response to mistletoe extracts. This result was enough for the producers of herbal products to get their bandwagons rolling, bandwagons loaded up with mistletoe extracts sporting intriguing names such as Iscador, Eurixor, or Helixor. Unfortunately, human trials have not born out the early optimism and there is no evidence from properly controlled trials that such products have a beneficial effect on cancer. There is, however, plenty of evidence that they don't. It seems that mistletoe's magic is limited to enticing people to express their affections for each other. And that's nothing to scoff at.

Now, on to Santa's reindeer. And let me warn you, this discussion is going to take us pretty far afield. As far as the snow-covered fields of Lapland and Siberia, in fact, where the legend of Santa Claus may have originated. Hundreds of years ago, the inhabitants of these barren lands tried to domesticate reindeer. But herding these animals was not a simple task, at least not until the herders found a most unusual ally in the form of a little red-and-white mushroom. Reindeer, it seems, just loved *Amanita muscaria*, or "fly agaric," as it is better known. They could be led by the nose, as it were, just by sprinkling pieces of the mushroom in front of them. Why the animals were so fond of these fungi was a question that must have occurred to many curious people. To find the answer, all they had to do was taste the mushroom. A blissful euphoria would have quickly permeated their bodies; vivid visions, perhaps even of sugarplums, would have danced before their eyes; and heavenly sounds would have resonated in their ears. Unless they overindulged. Then there would have been dead silence.

Amanita muscaria contains the natural hallucinogens muscamol and ibotenic acid, along with the potent toxin muscarine.

Amanita is not the "magic mushroom" that people grow illegally in basements or crawl on their knees to find in the forests of the Northwest. Those are *Psilocybe* mushrooms, which contain the hallucinogen psilocybin. Psilocybin can also cause problems, but it is not nearly as toxic as muscarine. Fly agaric is indeed dangerous stuff to mess with. Early Laplanders and Siberians probably took their chances with the fungus in order to while away the long, dark, dreary nights. This practice no longer occurs, as the "happiness mushroom" has been replaced by television, which in most cases provides a less toxic form of entertainment.

Amanita ingestion was never widespread because the fungi were relatively rare. It was usually reserved for special occasions, like Christmas. The well-to-do would indulge and often emerge from the celebrations to relieve themselves in the snow. This practice had the effect of attracting reindeer, which would vigorously lap up the melted snow. Impoverished natives noted this reindeer behavior and jumped on an easy way to brighten their lives. They gathered the yellow-tinged snow, melted it, and partook of the unusual beverage. Obviously, some of the mushrooms' active ingredients were eliminated in the urine.

The powerful hallucinogenic properties of *Amanita muscaria* led to its being used by shamans in quasi-religious ceremonies. Shamans were healers, spiritual leaders, and psychologists all rolled into one. They would often consume the hallucinogenic mushroom to enter a trance-like state of heightened awareness that appeared more conducive to physical and spiritual healing. Ceremonies were heralded by a stick placed upright through the smoke hole of the shaman's hut, a hole through which the spirit of the shaman supposedly exited and entered. It was also common for shamans to offer samples of fly agaric to their flocks, samples that were carried in a leather sack. According to some accounts, the association between these rituals and the mushroom was so strong that shamans decorated their fur clothing with red and white, the colors of the mushroom. So there you have it: a cheerful man dressed in red-and-white fur bearing a sack of presents. Add to this picture the imagery of the smoke hole and the hallucinating reindeer and you've got yourself a jolly old Santa and his "flying" entourage. As Ripley would say, believe it or not.

In this case, maybe we'll set scientific scrutiny aside and go for the "not." Why bring Santa down to earth with a dose of science? Let's let him fly on the wings of imagination and spread his message of happiness and good cheer far and wide. If it's reality we're after, why not just look into children's eyes as they anticipate a visit from the jolly old elf. The laughter in them is as real as can be.

There's No Viagra in Niagara

I suspect that milking a swamp rabbit is a pretty tough task. Milking the public, on the other hand, is quite easy. Just ask Silent George of Shawneetown, Illinois. Actually, you can't. The clever conman is no longer with us. But early last century he really did have an ingenious little scam going. George stripped the labels off cans of condensed milk, spray-painted the cans gold, and affixed them with new labels declaring the contents to be "Swamp Rabbit Milk." This milk was described as a "balanced formula for unbalanced people, rich in vitamins J, U, M and P." The name of the product was deliberately chosen to conjure up mental images of the legendary mating habit of rabbits. Swamp Rabbit Milk, you see, was promoted as an aphrodisiac. There were testimonials galore about its effectiveness despite the fact that vitamins J, U, M, and P do not exist. The product worked for the same reason that other supposed aphrodisiacs work. It stimulated the most effective sex organ of them all: the one between our ears.

Humans are absolutely passionate when it comes to searching for passion. The variety of substances that have been tried over the years to provoke sexual desire is truly astounding. Bird's nest soup and ginseng were ancient Chinese favorites, while the Kama Sutra, compiled between 100 and 300 A.D., recommended an elixir made of honey, milk, licorice, and fennel. Pliny, the Roman philosopher, believed that consuming a lizard drowned in urine had an aphrodisiac effect on the person who donated the fluid. People who were adverse to lizard consumption could resort to dining on the right lobe of a vulture's lung. Foods that resembled appropriate parts of the human anatomy were especially prized. Avocados were a turn-on for the Aztecs and famed European herbalist Nicolas Culpepper recommended asparagus for stirring up bodily lust. Sixteenth-century Italian physician Dr. Leonardo Fioravanti prescribed a tonic of nuts and cinnamon sticks for men who needed a little help. Casanova downed oysters regularly, supposedly because of their resemblance to the female private parts. Rhinoceros horn, nutmeg, truffles, deer antlers, elk horns, seal penises, ground goat testicles (only the left one), M&M's (the green ones), and, of course, wild oats have all been touted for boosting the libido.

You might think that these attempts to aid sexual performance owe their popularity to a less scientifically sophisticated bygone era, but you would be wrong. There are more purported aphrodisiacs available today than at any other time in our history. Believe it or not, a few may have some interesting physiological properties. Niagara, a beverage created in Sweden and promoted as "Romance in a Bottle," has caused quite a stir in the US. The American rights belong to Lari Williams, an Arkansas café owner who managed to excite about 60,000 people into ordering the blue bottles after she made a TV appearance on *Good*

Morning America. She also managed to excite the Pfizer pharmaceutical company, makers of the drug Viagra, which is used to treat erectile difficulties. Williams, the company said, was unfairly capitalizing on the success of its product and promptly took her to court. Pfizer wants the name "Niagara" changed, as the beverage has nothing in common with its own product.

Indeed, there is no Viagra in Niagara. The drink is sold as a dietary supplement and contains extracts of damiana (with the interesting botanical name *Turnera aphrodisiaca*), ginseng, and schizandra, a Chinese berry. These plants have a history of aphrodisiac hype, but there are no scientific studies to corroborate their effectiveness. Caffeine is also included and may account for some of the stimulation people claim to feel. Pfizer understandably frowns on any "aphrodisiac" association with Viagra, as the drug does not stimulate the sex drive; it only enhances performance by increasing blood flow to the genital area. Viagra is expensive and many men wonder about the existence of "natural" products that can be of similar help.

Corynanthe yohimbe is a tree that grows in West Africa. Various preparations of its bark have long been used to treat sexual dysfunction and modern research bears out the folklore, at least to a degree. Yohimbine, found in the bark, blocks alpha-2 adrenergic receptors on nerve cells—a complicated way of saying it can increase blood flow to the penis. An analysis of yohimbine studies shows that 30 percent of men using it achieved satisfactory erections compared with 14 percent using a placebo. This result may sound better than it really is; the data actually show that about six men would have to be treated for one to be helped significantly. (For Viagra, the ratio is 2 to 1.) Then, of course, there is the usual problem with herbal remedies—namely, lack of regulation to ensure that the product really contains what it is supposed to contain. Yohimbine hydrochloride is available as a prescription drug and has a greater potential for success. Interestingly, in rats as well as in Nile crocodiles, yohimbine seems to increase not only performance but also the libido. Once again there is a lack of human evidence in this area, but it is well known that yohimbine can cause anxiety, heart palpitations, and sleeplessness. In fact, Consumers Union has listed yohimbine as one of the dietary supplements that should never be taken.

L-arginine, a commonly occurring amino acid, is also receiving attention as a performance enhancer. It is the raw material our body uses to make nitric oxide, a compound that produces smooth muscle relaxation. In fact, Viagra works by blocking an enzyme that breaks down nitric oxide. Again, there is a lack of human data on arginine, but in theory it should have an effect. Jed Kaminetsky, a New York urologist, sure thinks so. He has formulated Dr. K's Dream Cream for women, which he says increases blood flow to the essential area and increases pleasure.

If none of these options appeal, you might consider a trip to Scotland. The tourist board there has launched a campaign that features "aphrodisiac haggis." Heavenly Haggis is made with apricots, figs, and pine nuts in addition to the usual less appetizing components, and Hot 'n Horny Devil Haggis turns you on with chili and Cajun spices. Alternatively, you might want to check out the Aphrodisiac Restaurant in Singapore, which has special menus tailored to the specific needs of men and women. In Italy, people travel from all over to Cascina Orlowsky in Piedmont to feast on *frittata afrodisica,* or aphrodisiac omelet. But the hottest spot these days seems to be Tantra in Miami's South Beach. The signature Tantra plate has oysters and barbecued eel and can be washed down with a shot of rum, vodka, and peach schnapps fortified with a blend of aphrodisiac herbs. I think the elixir would work even better if the folks at Tantra added some Swamp Rabbit Milk.

THE WONDERS OF POLYESTER

I've always had a special fondness for polyester. That's because my first-ever summer job was knitting with it. Don't go conjuring up any mental images of rocking chairs and knitting needles. My job was to look after the mechanized knitting machines that wove polyester fiber into fabric for the "leisure suits" that were destined to become the butt of many a joke but in the 1960s and 1970s were the height of fashion. Who can forget John Travolta spinning around the dance floor in *Saturday Night Fever* in his white polyester suit? He didn't even have to worry about perspiration stains, as polyester was remarkably easy to clean. Nor did he have to be concerned about creases, as polyester was the first real "permanent-press" fabric. Good thing, too; if you tried to take an iron to the material, it melted. But that property eventually turned out to be a useful one, enabling items made of polyester to be recycled.

Polyester was the world's first truly synthetic fiber. Credit for its discovery goes to Wallace Carothers and his assistant, Julian Hill, both of whom worked at the DuPont chemical company. Carothers is best known for his discovery of nylon, but his work on polyester actually predates this remarkable invention. In the early 1900s, a great controversy enveloped the chemical research community. To account for the physical properties of substances such as proteins and cellulose, Germany's Hermann Staudinger had suggested that small molecules (called "monomers") could be joined together into long chains called "polymers." Others maintained that polymers were nothing more than monomers that were somehow closely packed together. Carothers decided to solve the mystery by attempting to lace together small molecules by known chemical reactions to see if a polymer would result. Knowing that alcohols and carboxylic acids formed compounds called "esters," he used molecules that had alcohol and acid functions on both ends to create "polyesters." In 1931 Carothers published a paper in *Chemical Reviews* describing his results and essentially creating the field of polymer chemistry. Just the year before, Julian Hill had made the critical observation that melted polyester could be drawn into fibers and, perhaps even more important, pulling the cold fibers under tension hardened them significantly. Subjecting a fiber to such stress causes the long polyester molecules to orient themselves along the axis of the filament, leading to enhanced strength. In much the same way, intertwining steel strands leads to steel cable.

I didn't know anything about oriented polymers when I watched my knitting machines spewing out polyester fabric. But I do remember an interesting feature of the job. Quite often I would have to tear loose polyester thread from the material, which turned out to be quite easy when done quickly, but much tougher when done slowly. Increasing the tension on the thread slowly caused the polyester molecules to orient themselves further and toughened the strands. After this summer job I didn't give polyesters much thought until 1969, when, along with millions around the world, I watched one of humankind's greatest technological achievements as Neil Armstrong guided the lunar lander to a successful touchdown on the moon. You may recall that the lander was wrapped in a lustrous gold material. That material was Mylar, a sheet of polyester coated with a thin layer of gold, designed to protect the lander from the powerful rays of the sun. That's how I learned that polyester could be not only drawn into threads but also melted into superstrong flexible sheets. You may have encountered this same material in the form of emergency blankets that fold up into a pocket-sized pack and have a remarkable ability for retaining heat. The same polyester material is used as the basis of audio and video-tapes, and where would civilization be without those?

Given this background, you'll understand why I was immediately drawn to the movie *Polyester*, which debuted in 1981. I thought I was going to see a movie about chemistry. Hardly. The film was a mocking view of suburban life and starred drag queen Divine as beleaguered housewife Francine Fishpaw. There were numerous shots of consumer products, the acquisition of which defined "success" at the time. Many of these products were supposedly made of polyester, which in the movie came to represent the artificiality of suburbia. This metaphor did not reflect well on this fascinating material. Nevertheless, *Polyester* did have an interesting chemical connection: it was the first film shot in Odorama! Everyone was given a scratch-and-sniff card upon entering the theater and guided by instructions on the screen about when to scratch, audience members experienced the same invigorating fragrances as the characters in the film. There were oven fumes, pizza, "new car," and sweaty gym sock smells, and I'll leave it to you to guess which odor filled the theater when the number 2 appeared on the screen. And guess how these scratch-and-sniff cards worked? The smells were encapsulated in tiny plastic beads, made, coincidentally, of polyester!

Polyester has come a long way since *Polyester*. Those large ubiquitous pop bottles are made of it—thankfully. Large soda bottles made of glass can act like bombs. The glass has to be thick to withstand the 4 atmospheres pressure; if the bottle is dropped, shards can fly all over the place. Not so with polyester. Again, the key is stretching the plastic in both directions to increase its strength by reorienting the molecules. Beer, though, presented a problem. Plastic bottles were deemed desirable because they are unbreakable, lighter to transport, and allow the beer to be cooled more quickly. But in the case of beer, preventing entry of oxygen and loss of carbon dioxide through the plastic is more critical than with soft drinks. This problem has been solved by a design that adds thin nylon barrier layers to the polyester. Apparently these layers are readily separated during recycling and are not an impediment to the process. In 2000, the Miller Brewing Company became the first brewer in the US to make beer available in plastic bottles.

Recycling is one of the great advantages of polyester bottles. They can be melted down and extruded into fibers that are far superior to those that made up the leisure suits of the 1970s. Today's "microfiber" is silky and can be made into fleece that is not only warm and cozy but attractive as well. In fact, my favorite sweater is made of recycled polyester bottles. Wearing it not only represents a neat bit of chemistry but also brings back memories of a summer job and a stinky movie that introduced me to the wonders of polyester.

PAY DIRT!

Chemists have a special appreciation for gold. After all, the history of our science is intertwined with that of alchemy, the age-old quest to convert "base" metals into gold. By the nineteenth century it became obvious that this was not going to happen; gold was an element and could not be made from other substances. If you wanted gold, you had to find it. Since it is an extremely unreactive element, it does occur in nature in its native, or uncombined, state. That's why you can actually find pieces of pure gold—if you know where to look. Like in the dirt at the bottom of some Alaskan streams. I know, because I found some there.

I had looked forward to visiting Alaska because one of my favorite lectures deals with the chemistry and history of gold. On numerous occasions I had described to students how Joe Juneau, a French-Canadian prospector, discovered gold in

Alaska in 1880 by panning in a stream near the little town that would eventually bear his name. Now I had the chance to stand beside that very stream and pan for gold myself. Funny how talking about something is so much easier than doing it. Of course, I knew the basic concept behind panning. Gold is usually embedded in quartz and is liberated when rapidly flowing water erodes the rocks. Being very dense, it sinks and mixes with the stones and silt at the bottom of the stream. The idea, then, is to pan out some of this sludge and slowly swirl it with water, allowing the heavy gold particles to accumulate at the bottom of the pan while the lighter mud is poured off. It sounds pretty straightforward. But it took at least a half hour of swirling, pouring, and swearing until the first golden pieces began to glisten in the pan. I knew they would appear eventually, since we had been "guaranteed" to find gold. Truth be told, the pans we were given had been preloaded with dirt from some other location, as the original Juneau site has long since been exhausted.

Gold has always represented wealth and security. When the Israelites feared that Moses would not return from Mount Sinai, they found comfort in the building of the golden calf. The Egyptians, for their part, often buried pharaohs in golden caskets surrounded by gold artifacts to ensure prosperity in the next world. But perhaps it is the story of Jason and the Golden Fleece that best epitomizes humankind's desire to go to the ends of the earth in search of gold.

According to Greek mythology, Jason and the Argonauts braved adversity to bring the fleece to Greece from the kingdom of Colchis. The story probably reflected the fact that since antiquity, sheepskins have been used in what is today the post-Soviet Republic of Georgia to filter gold from rivers. How did gold get into rivers in the first place? According to legend, King Midas was granted a wish by the god Dionysus. He wished that everything he touched would turn to gold. Midas soon realized his folly as his water, food, and women began to glisten. He begged to be released from this curse and was told to bathe in a nearby river, which may explain why we still find gold in rivers. Or perhaps flowing rivers just happen to erode gold-bearing quartz rocks, leaving tiny pellets of the metal in the riverbeds.

It was those glistening pellets that attracted prospective prospectors from around the world to Alaska and later, in 1896, to the Yukon, hoping to hit "pay dirt." For most, the effort did not "pan out." After sailing to Alaska on some rickety vessel, they landed in Skagway, hundreds of miles from the gold fields. Battling ice and snow, they struggled over the famed Chilkoot Pass to Lake Bennet, from which Dawson City and its gold fields were within reach by boat—which the prospectors had to build themselves. When the adventurers got to Dawson, they discovered that the streets were not paved with gold. In fact, there were no streets at all. But for those willing to work hard under unimaginably harsh conditions, there was indeed gold in them thar hills. A few of the gold rushers accumulated incredible wealth, but for most who made the long, tortuous, inland trek, the experience was a bitter one.

Today most of the world's gold is mined in South Africa and Russia. In one of the processes used to extract gold, gold-bearing rock is crushed and then washed over copper plates, the surfaces of which are covered with mercury. Gold dissolves in mercury (as anyone who has been unfortunate enough to allow mercury to come into contact with their gold jewelry knows) and the resulting gold amalgam is scraped off and distilled. The mercury evaporates and the pure gold is left behind. About 4 tons of rock is needed to eventually yield 1 ounce of gold.

The pure metal is referred to as 24 karat, while 18 karat represents an alloy of 75 percent gold and 25 percent other metals. When these metals are nickel, copper, and zinc, white gold results; when the metal is silver, the product is the familiar yellow gold. The metal can even be given a reddish or greenish tinge by the judicious use of silver, copper, or zinc. A thin layer of gold is highly reflective and can be used on windows to save on air conditioning costs. It is such an excellent conductor of electricity that it is used extensively in computer circuits. In fact, a new industry has recently arisen: recovering gold from computer scrap. One ton of computer rubble can yield 2 pounds of gold. The greatest gold reserve in the world is actually seawater. Although the gold concentration in seawater is very small, the massive amount of water in the world's oceans means that the total dissolved gold is more than what has been extracted from all other sources throughout our history. The only problem is that the technology to extract it in an economic fashion does not exist.

Gold is the most malleable of all metals; an ingot the size of a matchbox can be beaten into a sheet 1/10,000 of a millimeter thick, making it large enough to cover a tennis court. Knowing this, I shouldn't have been surprised when, after my gold-panning adventure, I walked into a souvenir shop in Juneau and was greeted by a large display of little vials filled with flakes of real gold. Each one contained far more gold than the paltry specks I had managed to scrape together. The cost? Ninety-nine cents each. Compare that with the $39 per person for a five-minute bus ride and half hour of sifting mud. Obviously there still is gold to be found in Juneau. You just have to mine it from tourists' pockets.

A Galling Tale of Vitriol and Murder

Even the world's greatest forgery experts were forced to admit that Mark Hofmann's work had been superb. In fact, many of them had been fooled by the man who appeared to have had amazing luck stumbling across rare historical documents penned by the likes of George Washington, Daniel Boone, and Joseph Smith, the founder of the Mormons' Church of Jesus Christ of Latter Day Saints. In 1985, a letter supposedly written by Smith and "discovered" by Hofmann turned the Mormon community upside down because it seemed to undermine some of the basic tenets of the Church. The elders argued that the document could not be authentic and that Hofmann had a lot of gall claiming it was.

Actually, they were right. Hofmann had pulled off the forgery with gall. Iron gall. More specifically, ink made from iron gall. And therein lies a story that begins with an ancient Roman chemical marvel, takes us through Leonardo da Vinci's fading notebooks, explores Hofmann's clever forgeries, and ends with an insight into the mysteries of colon cancer.

Way back in the first century A.D., Gaius Plinius Secundus, better known as Pliny the Elder, entertained his fellow Romans with what seemed to be a magic trick. He dipped a strip of papyrus into one colorless liquid, then into another, and to his audience's amazement, the papyrus immediately turned black. Pliny had chanced upon a chemical reaction that would eventually revolutionize writing. He had made iron-gall ink.

So what are "galls" and where are they to be found? These galls have nothing to do with bladders and stones; they have to do with oak trees. Galls are little spherical growths that form when insects such as wasps, aphids, or flies puncture the twigs of oak trees to lay their eggs inside. The tree forms galls in an attempt to isolate the parasites by encapsulating them. Oak galls are particularly rich in compounds called "tannins," which have antiparasitic properties. Tannins can be extracted from the galls with hot water, and one tannin in particular, gallotannic acid, plays a central role in our story. When a solution of gallotannic acid is allowed to stand, it readily becomes contaminated by molds from the air, which trigger a fermentation as they release certain enzymes. During the fermentation, gallotannic acid breaks down and liberates gallic acid. Fermentation is not critical; gallic acid can be directly freed from gallotannic acid if the original extraction is done with an acidic solvent such as wine. The latter was probably Pliny's technique.

Gallic acid forms a dark blue-black complex upon contact with ferrous sulfate. This substance appears as crystals on stalagmites in caves or on the walls of iron mines and was well known in antiquity as "vitriol." Pliny's trick amounted to nothing more than dipping his papyrus into a solution of gallic acid first and then into one of vitriol. But it wasn't until the Middle Ages that the product of this "magic trick" took on real importance as iron-gall ink.

Ink, of course, was not a new concept. Ancient Egyptians had long before used carbon in the form of soot to make markings on papyrus and parchment. Stirred into water and mixed with plant gums for adhesion, such carbon-based inks served well. But they also presented a problem. The carbon was deposited on the surface of the substrate and often smudged. It was also readily removed, posing a problem for documents that were supposed to be permanent. Would-be inventors pursued the elusive "permanent ink" until sometime around the tenth century Pliny's iron-gall ink reappeared, destined to fill ink bottles for hundreds of years.

This ink was highly satisfactory, as it actually became more permanent with time. Ferrous sulfate forms a soluble complex with gallic acid and readily penetrates the substrate. Then the ferrous iron is converted to the ferric form by reaction with oxygen in the air. The complex of ferric iron with gallic acid is insoluble and forms a black deposit wherever the original ink solution has penetrated. Iron-gall ink was so effective that collecting and selling oak galls became a lucrative business. Especially prized were the galls that had no holes, indicating that the insects inside had not yet managed to escape. Such galls were thought to have a higher content of gallic acid. Some scoundrels went so far as to fill in the holes in order to command a higher price. The required vitriol was usually produced from the fluid that trickled out of the walls of iron mines. It was collected in barrels, where crystals of ferrous sulfate readily formed on ropes immersed in the liquid.

The popularity of iron-gall ink was astounding, especially after it was discovered that adding gum arabic, an exudate of the acacia tree, allowed for a smooth transfer to the substrate. Documents ranging from da Vinci's notebooks to the US Constitution were written with this ink. And these historic relics are still with us. But a frightening question has cropped up: For how long? As conservators have discovered, residual ferrous ions in the ink serve as excellent catalysts for the production of free radicals from oxygen. These highly reactive species destroy parchment and paper. In fact, many valuable manuscripts produced with iron-gall ink are already filled with pinholes. A frantic search for a solution to this problem has ensued and a potential answer has emerged from, of all places, the medical literature. A current theory holds that free radicals produced by ferrous ions play a role in triggering colon cancer. This hypothesis has spawned research into the anticancer properties of phytate (inositol hexaphosphate), a chemical found in wheat bran, oat bran, and soybeans. Phytate is known to bind and deactivate ferrous ions. Preliminary experiments have shown that treating old manuscripts with a solution of phytate may prevent the deterioration process.

Mark Hofmann had studied the properties of iron-gall ink in detail. He knew that the ink "aged" and changed color upon exposure to oxygen. To mimic this process he treated his ink with a potent source of oxygen, namely hydrogen peroxide. This technique was part of his downfall. The peroxide cracked the gum arabic, giving it an alligator skin–like appearance, which is uncharacteristic of old documents. Furthermore, when the experts studied how the ferrous ions migrated from the ink through the paper or parchment, they concluded that the ink had been recently applied. Superforger Hofmann was sentenced to life in prison. That may sound like a harsh punishment for forgery. But apparently Hofmann's chemical skills extended beyond making iron-gall ink. He had murdered two potential investors with a homemade bomb to prevent one of his forgery schemes from being exposed.

The Right Alchemy: Harry Potter and Nicolas Flamel

"Daddy, when are you finally going to write about Nicolas Flamel?" I've been hearing that question for a while now from my daughter Rachel, who is addicted to everything connected to Harry Potter. For those of you who have been hibernating for several years and don't know the plot of the first tale in the Potter series, Flamel is the mastermind behind the remarkable philosopher's stone, which Harry must keep out of the clutches of the evil wizard Voldemort. The time has now come to reveal the truth about Nicolas Flamel. So gather your children around, because this story is for them. But not only for them. The adults may learn something too.

Once upon a time, about 400 years ago, there lived a cobbler in Bologna named Vincenzo Cascariolo. But this cobbler was interested in more than just making shoes last longer; he was determined to make people last longer! How? By finding the secret of the philosopher's stone. Vincenzo had to find a way to make the legendary magical rock that would not only turn base metals to gold but also yield the elixir of life. To do this, the Italian cobbler had to dabble in alchemy.

Cascariolo's fascinating quest already had a rich history. Gold, because it did not corrode and seemed to last forever, had long been revered as the most special of metals. If only they could discover how gold was created, the alchemists philosophized, they would become wealthy beyond their wildest dreams. Furthermore, they would have a very long time to spend that wealth. Whatever process was used to make gold immortal could likely be applied to humans as well.

Today, the very idea of "making gold" seems outrageous to a scientist. After all, we know that gold is an element and therefore can be neither broken down into simpler substances nor

formed from simpler substances. But a clear understanding of the existence of the elements and their ability to combine and form everything that exists in the world dates back only a few hundred years. The effort expended by alchemists to make gold may seem like futile bumbling in the light of current knowledge, but it was quite reasonable in the context of the times.

Alchemists saw iron become rust, silver tarnish, copper ooze from certain minerals when heated, and caterpillars change into butterflies. Why, then, couldn't some substance be "transmuted" into gold? Mercury, for example, looked metallic but wasn't the right color or consistency. Sulfur, on the other hand, was solid and was yellow. It certainly seemed conceivable that mixing mercury with sulfur under the right conditions would yield gold. Alas, it did not. The alchemists carefully documented their efforts, often using elaborate codes to ensure that whatever secret they chanced upon could not be stolen. One of the first to experiment with mercury and sulfur was an eighth-century Arabian alchemist named Jabir ibn Hayyan. He recorded his experiments in a code so confusing that it still hasn't been figured out. His notes look like "gibberish," a term we often use today without realizing that it derives from the name *Jabir*.

Since the alchemists repeatedly failed to make gold, they became convinced that something essential was missing from their recipes. Then, on April 25, 1382, that critical ingredient was discovered. It was the philosopher's stone, the key to transmutation. The discoverer was none other than the fabled Nicolas Flamel. Or so it was claimed in the books that lay open in front of Vincenzo Cascariolo as he sought to make the stone that had apparently eluded all others for 300 years. Unfortunately, Flamel's writings were very difficult to decipher. But Cascariolo seemed to be on the right track when he heated a mixture of barite (a commonly found mineral), powdered coal, and iron. To his amazement, he produced a substance that glowed in the dark! And once the glow faded, the amazing material could be reenergized just by exposing it to the sun. The Italian alchemist had discovered a way to capture the sun's rays! He thought he had surely taken the critical step toward making the philosopher's stone. In reality his concoction would never yield gold, but Cascariolo had stumbled onto the first glow-in-the-dark substance. He had converted barium sulfate (barite) to barium sulfide, a "phosphorescent" material.

Why was Cascariolo not able to reproduce Flamel's stone? Because Nicolas Flamel never made it in the first place. Oh, Flamel was a real person, all right. But he wasn't an alchemist. Flamel was a notary who lived in the fourteenth century and made a lot of money speculating on real estate. He and his wife, Pernelle, were kind people and used their wealth to establish hospitals and homes for the poor. Their seemingly bottomless pockets probably gave rise to the legend of the philosopher's stone. And a legend it was, as the books that were supposedly written by Flamel did not appear until the sixteenth century! Of course, some would argue that if Flamel had really discovered the stone, he would have been able to make himself immortal. If so, he could have written the books hundreds of years after his supposed death.

The facts would seem to argue against this possibility. Nicolas Flamel's gravestone is in plain view in the Musée de Cluny in Paris, France. But then again, maybe Nicolas faked his death and that of his wife in order to hide his success at having created the marvelous stone. Indeed, in the eighteenth century, some Parisians swore that they had seen Nicolas and Pernelle in attendance a performance at the famed Opera House. I don't know about that sighting, but I do know that today in Paris you can find rue Nicolas Flamel, a street that intersects with rue Pernelle. Nearby is one of Paris's most famous restaurants, the Auberge Nicolas Flamel. It is said to be housed in the oldest building in Paris, one that was originally built by the legendary Flamel to house the poor.

I plan to take Rachel there one day. And I'll raise a glass of wine in honor of the philosopher's stone and J. K. Rowling, author of the Harry Potter books. Why? Because while the stone is mythical, the concept is real. We call it "catalysis." A catalyst is a substance that speeds up the conversion of one substance into another without being consumed, like the yeast that converts grape juice into wine. Or the iron that Vincenzo Cascariolo used to convert barium sulfate to barium sulfide, briefly allowing him to bask in the light of what he thought was the philosopher's stone.

THE SWEDISH BLONDS WHO TURNED GREEN

It's obvious that there is a chemical problem afoot when Swedish blonds start sporting green hair after they shampoo. Health officials addressed the crisis quickly and found the problem. Acid rain had acidified well water in certain areas and caused copper plumbing to corrode. The result was that copper compounds, which tend to be green or blue, were dissolving in water and being deposited on hair. Actually, you don't have to travel to Sweden to observe this phenomenon. Just look at any copper roof. Over time, the copper reacts with acid rain to form copper carbonate and copper sulfate, which leach out and give rise to telltale greenish streaks on the masonry below the roof.

Why are we investigating the problems of Swedish blonds and copper roofs? Because of two items that recently appeared in my mailbox. One was a copper heating element that had been removed from an electric kettle, and the other was a bracelet fashioned out of copper wire. In the first case, the tin coating had worn away and there was concern about a possible health risk due to leaching copper. In the second case, the fledgling artisan wanted to know if the homemade bracelet would be as effective as the commercially available Tibetan medicine bracelet, which claimed to help arthritis. It also promised to bring "harmony and mental clarity to the wearer."

Let's look at the leaching problem first. And let's start in an unusual place: the high seas. During the early days of sailing vessels, ships' bottoms frequently became encrusted with all sorts of foreign materials, particularly barnacles. This was a problem because the resulting drag would slow the ships down. When copper sheeting became available in the late eighteenth century, shipbuilders began to cover wooden hulls with it, hoping that barnacles would not stick to the smooth surface. They didn't. But not because they couldn't get a foothold. This distinction became clear when iron hulls were introduced. Iron hulls had the same encrustation problems as wooden hulls despite their smooth surfaces. Copper, therefore, had some special property, and shipbuilders began to cover the iron hulls with copper sheeting as well. They were in for a surprise. Not only was the antifouling effect reduced, but the iron hulls also began to corrode!

Here's the explanation. Copper reacts with water, loses electrons, and forms positively charged, soluble copper ions. This reaction is very limited; nevertheless, any metallic copper that is in contact with water will have some copper ions in its vicinity. Iron is chemically more reactive than copper. In simple terms this means that it will donate electrons to copper ions, forming positive iron ions. Basically, this is the rusting process: hard metallic iron is converted to soft ionic iron compounds. Now we know why the hulls rusted. But why was the antibarnacle effect reduced? Because the protection had resulted from the toxicity of copper ions. When the concentration of these ions around the hull was reduced due to reaction with the iron, barnacles frolicked. Builders eventually solved the problem by covering iron hulls first with a layer of wood and then with a layer of copper. The most famous ship to make use of this technology was the *Cutty Sark,* launched in 1869. It was specifically built for speed; its owners aimed to win the annual race to bring back the first shipment of newly grown tea from China. Today, the *Cutty Sark,* moored in England, is the world's only surviving "tea clipper." Its copper-coated hull is clearly visible.

If copper ions are toxic to marine organisms at such low concentrations, could they be toxic to humans also? In higher concentrations, undoubtedly. This toxicity was clearly demonstrated by Pierre-Desire Moreau, a French herbalist in the 1800s. Authorities became suspicious when his second wife died after persistent vomiting—exactly the same symptom that wife number one had exhibited before her death. An investigation revealed large amounts of copper in her body. The first wife was exhumed and copper was found in her body as well. Both had been poisoned with copper sulfate. Moreau defended himself by saying that he had been unhappy and just lost his head. The judges were unimpressed. On October 14, 1874, he really did lose his head on the guillotine.

So it is evident that large doses of copper can be dangerous. But what about the small amounts we may be exposed to from copper pipes or heating elements? The mass of copper that dissolves depends on the acidity of the water and the temperature. Soft water (low in mineral content) tends to be acidic and will leach copper, particularly when it is hot. That's why you will sometimes see greenish stains in sinks and bathtubs. Hard water does not present as much of a problem. In any case, most municipalities adjust the acidity of water by adding a base to make it neutral or slightly alkaline to reduce corrosion.

Vomiting, diarrhea, and stomach cramps have been traced to high levels of copper in water. In one case, carbonated soft drinks (which are quite acidic) from a machine with copper plumbing caused the problem. In another, treated water was accidentally acidified. But such cases are extremely rare. Actually, copper is an essential element for health and we normally take in about 2 to 5 milligrams per day in our food and water. At a pH greater than 6.5 (which is the case for tap water), the solubility of copper is minimal. The amount that would leach out from an exposed element into a full kettle would be less than 1 milligram. This is not significant.

But what about the copper bracelet? In this instance, the argument is that small amounts of copper are leached out by acidic sweat and contribute to an increase in concentration of some copper-containing enzymes that may play a role in alleviating the symptoms of arthritis. The theory may sound a little far-fetched, but a study in Australia found that a copper bracelet loses about 1 milligram of copper a month and a placebo-controlled trial did show some benefit in alleviating arthritic pain. I remain skeptical. But I see no risk in a copper bracelet and am confident that a homemade version is as effective as the Tibetan one. (Of course, since I do not wear the Tibetan version, I may lack the mental clarity needed to make this judgment.) If you decide to give a copper bracelet a try, make sure to take it off when you shampoo your hair. That is, unless you want to suffer the same fate as the Swedish blonds.

SAVING LIVES WITH KEVLAR

One of the most unique clubs in the world is the Kevlar Survivors' Club. But I don't think you would want to join. The members are police officers who are alive today thanks to bulletproof vests made of the amazing material called "Kevlar." Interestingly enough, Kevlar didn't start its life as a material designed to save lives. It started out as a material designed to save time. Tires for racing cars were the intended targets. Steel-belted tires, of course, had long been in use and had proven their worth in terms of strength. But steel is heavy, and excess weight is always a problem in racing. So researchers at the DuPont chemical company looked for some material that could rival the strength of steel while being lighter. The catch: racing tires get pretty hot, so the material also had to stand up well to high temperatures.

The date was the early 1960s. The researcher in charge was polymer chemist Stephanie Kwolek. She was not exactly working in a void because DuPont was already a world leader in plastics research. It was here that Wallace Carothers had invented nylon and polyester back in the 1930s, two of the most successful plastics ever developed. DuPont did not rest on the laurels of nylon; the company was bent on improving the product. This process is par for the course in research; a discovery is made and research gears up to extend the work and use it as a springboard for further breakthroughs. In chemistry, this approach usually involves manipulating the structure of a molecule. For example, when a new drug is discovered, chemists start tinkering with its molecular structure in an attempt to come up with something that works better and has fewer side effects. This is just what Stephanie Kwolek did with nylon.

Nylon is a very flexible material. It is a polymer, a very large molecule that is constructed out of smaller building blocks called "monomers." Here's a crude analogy: Imagine you had little pieces of string that you tied together to make a long rope. The rope's properties would depend on the kind of string you used to make it as well as how you tied the pieces together. In the case of nylon the ties are strong, but the pieces of "string" are very flexible and the "rope" is easily twisted into any desired shape. Now imagine that instead of pieces of string, you used more rigid pipe cleaners. The pipe cleaners can be tied together in the same way as the string, but the final product will be a lot less flexible.

In a sense, this is exactly what the DuPont researchers did. Nylon had proved its worth in terms of the chemical bonds that linked the individual segments. These amide bonds were strong so that there was no need to change this particular aspect of the chemistry. But the individual segments had to be made more rigid. Kwolek and her crew decided to use a different kind of molecular structure for the individual units. In nylon, the basic fragments consist of carbon atoms joined in a row. Such straight chains of atoms are flexible. Previous work at DuPont had shown that if the monomers consisted of rings of carbon atoms instead of straight chains, the polymers they formed would be far more rigid. Such rings are referred to as "aromatic" rings, because the earliest samples studied—benzene for example—had distinctive smells. Stephanie Kwolek therefore focused her attention on "polyaramids."

Unfortunately, these substances did not melt easily and therefore the usual technique for converting them to fibers was not applicable. In general, fibers are formed by passing molten polymers through a spinneret, a device that resembles a showerhead. Kwolek had an idea: if she dissolved her polymer in a solvent, maybe the solution could be forced through a spinneret. It worked far better than she had hoped! Serendipitously, dissolving her aramid in a solvent had created a liquid-crystal solution. Such solutions can be described as a form of matter halfway between a liquid and a solid, and they feature an unusual degree of organization of the component molecules.

The polymer molecules basically lined up parallel to each other in the liquid crystal and were squeezed together into a close-packed arrangement as the solvent evaporated during spinning. Here's an analogy: Imagine taking cooked spaghetti, straightening out the strands, packing them together, and allowing them to dry. It'd be a lot harder to deform that dried bundle than a handful of limp spaghetti! Now imagine that the rigid spaghetti pieces were glued together at a few points. The glue would make for an even stronger structure. In Kevlar, certain hydrogen atoms on one molecule are attracted to oxygen atoms on an adjacent one. These "hydrogen bonds" constitute a "glue" that makes for a strong three-dimensional structure.

About $500 million and fifteen years later, DuPont had a viable commercial product. Kevlar turned out to be five times stronger than steel on a weight-per-weight basis. It was also flameproof and light. Unfortunately, its performance in racecar tires was not everything the researchers had hoped for. It certainly was strong enough, but its elastic characteristics—that is, its ability to deform and bounce back in response to pressure— were not ideal. "Aramid" tires, however, are great for those of us who do not need to take curves at 100 miles per hour.

It is the strength and light weight of Kevlar that has paid off in an amazing way. Kevlar threads can be woven into fabric, which can be used to make bulletproof vests. Such vests will stop most handgun bullets, although rounds from high-powered rifles will penetrate it. Kevlar's secret lies in its ability to deform and spread the impact over a large area. Officers shot while wearing Kevlar vests don't get away unscathed; there can be some pretty nasty bruising from the impact. But being bruised is a lot better than being dead. The ever-increasing membership in the Kevlar Survivors' Club (now over 2,500) is a strong testimonial to the chemical ingenuity of the group of DuPont chemists who twisted their minds to come up with a polymer that would not easily twist out of shape.

KISS AND MAKEUP

I don't know if Marilyn Monroe ever made it into heaven. St. Cyprian, a third-century Christian writer, surely would have argued that her chances were pretty slim. God looked unfavorably on women who showed up at the Pearly Gates wearing lipstick, he maintained, because humans were created in God's image and altering this image was sinful. The Creator, presumably, did not wear lipstick. But Queen Nefertiti, who ruled Egypt with her husband, Akhenaton, from 1379 to 1362 B.C., did and would undoubtedly have had a different view on the matter.

How do we know that an Egyptian queen who lived over 3,000 years ago used lipstick? It's pretty obvious if you visit the Egyptian Museum in Berlin, Germany. The museum's prime attraction is a bust of Nefertiti, one of the most celebrated ancient artifacts in existence. It was unearthed during a 1914 archeological dig sponsored by Dr. James Simon, a German merchant. The bust was smuggled out of Egypt and eventually donated to the Prussian State by Simon. In 1933 the Egyptian government demanded its return, but Adolf Hitler would not hear of it. The statue was a continuous delight to him, he claimed, and would be prominently displayed in its own room in a museum he planned to build. That room would be right next to the hall that would house the museum's centerpiece, a bust of the Fuhrer. Lucky for Nefertiti's statue, the shared space never materialized. Today Nefertiti rules the Egyptian Museum of Berlin, basking in the glow of a single spotlight in her private chamber. The limestone bust was beautifully painted by its sculptor, leaving no doubt that the ancient Egyptians used cosmetics for their eyes and lips.

What did they use? The dark makeup around the eyes was a blend of black lead sulfide and white lead carbonate. These substances were derived from galena and cerrussite, minerals mined around the Red Sea. We come by this knowledge thanks to the Egyptians' belief in an afterlife. Life, they thought, would somehow continue in the next world, including the usual human activities of eating and procreating. Accordingly, tombs were filled with food supplies and cosmetics. The Egyptians figured the need to be attractive and to seduce did not stop with death. Numerous cosmetic containers have been recovered from ancient Egyptian burial sites and their contents analyzed by modern chemical techniques. Lead compounds used in eye makeup have been definitively identified; they were mixed with fat (goose fat seems the most likely candidate) to facilitate spreading. But what Nefertiti and others may have used on their lips is more of a mystery. The best bet is that Egyptian women reddened their lips with clay tinted with iron oxide or cinnabar, a naturally occurring form of mercury sulfide. This practice could have led to unrecognized mercury poisoning. Luckily, Egyptian women may have had another option for coloring their lips with greater safety.

Though bug juice does not sound like an appetizing cosmetic, it may have served the purpose. A powerful red dye produced by squishing the wingless kermes insects that inhabit certain species of oak trees was well known at the time and certainly could have been used to dye lips. Some sources claim that the Egyptians used carmine, an extract of the cochineal insect, for this purpose. Historians would disagree, as these insects are native to South America and were not introduced into Europe until the sixteenth century. Without much doubt, though, carmine was the basis of the first modern lipstick, which dates back only to about 1915.

Perhaps we shouldn't be too surprised that lipstick is a relatively recent development. After all, there is a lot of sophisticated chemistry involved in its production. Just think of the requirements: the components cannot be toxic, cannot taste bad, and must not irritate or dry the lips. A lipstick must not melt in a purse; it must spread evenly and have good keeping qualities. Of course, the coloring agent cannot be water soluble, as even the slightest drool would result in streaks on the chin. These stringent conditions are met, not through random mixing of substances, but rather through application of scientific principles.

Oil extracted from castor beans and mixed with beeswax is an ideal base for lipstick. This combination, modified with carnauba or candelilla wax (both from plant sources), makes for a stiff "thixotropic" blend that softens when pressure is applied. The mix is somewhat sticky but the addition of isopropylmyristate reduces drag. Emollients such as cholesterol or lanolin from sheep's wool are added to prevent water loss and keep the lips moist. Butylated hydroxy toluene (BHT) or butylated hydroxy anisole (BHA) are preservatives that prevent the oils in the product from going rancid. You'll also find these chemicals in your cereal. Then, of course, there are the colors. Carmine is still used, but modern chemistry has made available a spectrum of other pigments. Most of these pigments are rendered insoluble in oil or water by reacting them with aluminum compounds to produce a "lake," which is then ground into very fine particles that can be suspended in the base.

"Color-changing" lipsticks use an alternate chemistry. One of the ingredients (e.g., eosin) is only lightly colored until it becomes a strong red upon reacting with proteins in the skin. In the tube, the lipstick is blue, colored with an aluminum lake. When applied to the lips, the red color develops and overpowers the blue shade—a bit of fun and some interesting chemistry!

Is there any health risk with lipstick? Of course, there are the usual inflammatory Web sites that claim women ingest 4 pounds of the stuff over a lifetime and denounce lipstick as "the most dangerous of all cosmetics." Absurd! Rare allergies to lipstick are possible. In one case, a young boy developed breathing problems after lipstick had been applied for his performance in a play. Later, his face swelled when he was kissed by his grand-mother.

The biggest risk facing lipstick wearers may be getting it on someone's collar. Lipstick stains can damage fabrics as well as lives. Rubbing alcohol and a prewash product such as Shout can usually get rid of the stain. Fixing lives damaged by the wrong shade of lipstick is tougher. President Kennedy might have known something about the latter. If St. Cyprian was right, Marilyn Monroe's trademark red lipstick may have kept her out of heaven. However, it is rumored to have gotten her into the White House.

HARD-WATER WOES

Baby Culligan's bottom was as soft as, well, a baby's bottom. Such soft skin was unusual in the early 1900s in Minnesota, where most babies suffered from the effects of diapers laundered in hard water. But young Culligan purred quietly in his comfortable diapers, washed in the soft water produced by the process his father had just invented. Emmett Culligan immediately recognized the market potential of catering to babies' bottoms and began to advertise his water-softening filter. Hard-water woes, he claimed, would become relics of the past. Complaints included a lack of soapsuds when washing as well as an unsightly gray film left on laundry, hair, and skin. To add insult to injury, people lucky enough to have hot water pipes often saw their water flow restricted as scale deposits built up inside the pipes.

Water is an excellent solvent, particularly if it is somewhat acidic. Rainwater fits that bill. Its acidity is due to carbonic acid, which forms when carbon dioxide from the air dissolves in rain. As this acidic water percolates through the ground, it dissolves a variety of minerals. Limestone, or calcium carbonate, reacts with the acidified water to produce soluble calcium bicarbonate. Similarly, dolomite, which contains magnesium carbonate, will produce soluble magnesium bicarbonate. These soluble forms of calcium and magnesium account for the "hardness" of water. Why hardness? Because it was long known that soapsuds were "hard" to produce in certain waters, a difficulty that was eventually traced to the presence of calcium and magnesium in the water. While soap is soluble in pure water, it produces an insoluble scum when it reacts with calcium or magnesium ions. This scum creates your classic bathtub ring! It has nothing to do with dirt. You can be immaculately clean and still produce the ring if you're using soap in hard water. And the water doesn't have to contain much calcium or magnesium. Over 120 parts per million and you've got hard water!

Soap scum can also deposit on the skin or hair and be very difficult to rinse off. Laundry presents an even greater problem. Not only is the scum hard to remove, but even small residues can leave white fabrics looking dirty. Furthermore, some soils combine with calcium or magnesium to form hard precipitates that get embedded in the laundry, making for an unpleasant scratchy texture. The advent of synthetic detergents solved many hard-water problems. Unlike soaps, these detergents did not form insoluble compounds with the hardness minerals. Still, the minerals in solution did impair the activity of the detergents, requiring that a larger quantity of them be used. Removing these minerals, or "softening" the water, was still desirable.

Culligan had been a young, successful real estate entrepreneur but lost his fortune after making some bad investments. He was looking around for another career when a childhood friend introduced him to a material called "zeolite." This unusual, naturally occurring rock, the friend said, had the ability to exchange calcium and magnesium ions for sodium ions. Culligan knew that many would-be inventors had tried unsuccessfully to soften water, and he quickly recognized the opportunity that was knocking. He took an empty coffee can, perforated the bottom, and filled it with zeolite. Then he passed water through this "filter" and was elated to see that soap now produced copious suds. Joy reigned supreme when Culligan found that his filter, which would eventually become saturated with the calcium and magnesium removed from the water, could be regenerated just by treating it with a concentrated salt solution. And the water-softening industry was born! Today in hard-water areas, many people find life worth living again after they hook up their washing machine's water supply to a water softener.

Hard water, though, can present an even more insidious problem. Have you ever noted what happens on the inside of an electric kettle? The elements become coated with a grayish scum. Heat actually reverses the reaction that caused the calcium and magnesium compounds to dissolve in water in the first place; both calcium and magnesium bicarbonate release carbon dioxide and turn into insoluble calcium and magnesium carbonate. In a kettle you can deal with this problem pretty readily. Just add an acid—vinegar, for example—and the deposits will dissolve. But inside a hot-water boiler or industrial pipes that carry hot water, the solution is not so simple. In fact, the problem is a significant one. In boilers and cooling systems, the scale that builds up does not transfer heat well, which results in energy costs that are increased by as much as 25 percent. Hot-water pipes can become clogged with scale, reducing water flow and requiring replacement. The annual cost of damage attributed to hard water in North America runs into the billions.

Industry uses a number of water-softening techniques, including Culligan-type ion-exchange filters. The traditional method, however, involves adding lime (calcium hydroxide) and washing soda (sodium carbonate) to the water to precipitate out the "hardness" minerals prior to use. Some industries claim that clamping magnets—often not much stronger than refrigerator magnets—around hot-water pipes reduces scaling. Numerous experiments have been conducted to validate this claim, but the results are inconclusive. Some studies show an effect; others do not. It seems that when magnetic softening does work, some other factor must come into play as well. The theory that has been put forward to explain the magnetic effect suggests that the magnetic field somehow interferes with crystal formation and that while scale still forms, the crystals of calcium and magnesium carbonate are less likely to stick to the wall of the pipe. In any case, the effect is not great; if it were, we would not be arguing about it.

While soft water is great for hot-water pipes, it isn't desirable for our internal plumbing. Drinking softened water unnecessarily increases sodium intake. Furthermore, calcium and magnesium are beneficial minerals. Some studies have even suggested that heart disease rates are lower in hard-water areas. That's a controversial claim, but there is no question that laundering in soft water does result in cleaner and softer fabrics. A cruise line recently capitalized on this well-established fact by advertising that it washes its sheets in ultrasoft water for the benefit of passengers who stay out in the sun too long. Shades of Emmett Culligan.

LOOK AT THOSE COW CHIPS FLY!

"Unfair!" screamed the spectators who ringed the field at an Iowa county fair. They had been patiently waiting for the epic confrontation between the state's best cow-chip slingers and anticipated a close battle. But they were disappointed. The newcomer's sleek cowpat easily outdistanced the reigning champion's bulky projectile. Had the rules, though, been stretched? To the spectators, the winning entry looked more like a wooden Frisbee than a cowpat. The new champion, however, insisted that there was as much poop in his entry as in any of the others. And he was right! What looked like a wooden disk was essentially a composite of soy protein and manure. It had been made by researchers at Iowa State University in response to a request by the Iowa Soybean Board to promote the notion of using soy protein as glue in the manufacture of construction materials.

Particleboard is a mainstay of modern construction. It is made by gluing wood chips together with a synthetic resin such as urea-formaldehyde. While it is clearly a useful material, there has been concern about the release of formaldehyde vapors, prompting research into alternative adhesives. Scientists at Iowa State University have found that glue made of soy protein mixed with phenol-formaldehyde resin does an excellent job without releasing any formaldehyde vapors. And, of course, the soy protein comes from a renewable resource! They also found that wood chips can be replaced with other forms of fiber, such as ground cornstalks, or, amazingly, cow manure. Perhaps we should not be surprised. After all, cows eat grass, which is a fibrous material. They process it in the same way that a wood chipper processes wood: long pieces go in and short shreds come out. Except as wild-card entries in cow-chip-throwing contests, we'll probably never see construction boards made only of manure and soy protein. But these materials can certainly play a role in replacing some of the traditional components of particleboard.

The idea of using plant products to formulate industrial materials is certainly not new. Henry Ford was intrigued by this idea way back in the 1920s. He had grown up on a Michigan farm and became greatly concerned when he saw farmers' fortunes decline during the Great Depression. Of course, he was also concerned that they would not be able to buy his cars and trucks. "If we want the farmer to become our customer, we must become his," Ford proclaimed. But how could the fledgling automobile industry use crops? To find out, Ford established a laboratory in Dearborn, Michigan, headed by Robert Boyer, his handpicked protégé. Ford suggested that soybeans, which were known to be rich in oil, protein, and fiber, should be explored as possible raw materials for automobile components. He turned out to be right!

By 1933 Boyer had developed an enamel paint for cars formulated with soy oil, and by 1935 he had found a way to make plastic from soy protein by subjecting it to high pressure. Soon the horn buttons, gear-shift knobs, and accelerator pedals in all Ford automobiles were fashioned out of soy plastic. Then Boyer came up with something even more impressive: soy-plastic panels to replace metal sheets! Ford, never adverse to publicity, had a field day with this invention. He would routinely entertain guests and reporters by flailing away with an axe at his personal vehicle, which had been fitted with a soy-plastic trunk lid. The panel withstood the barrage. Ford then invited onlookers to do the same to their own automobiles. There were no takers. With great glee, Henry would then dust off his suit, pointing out that it too was made from soy fiber.

In 1941 Ford unveiled a car whose whole body was plastic and predicted that its construction was the wave of the future. The plastic body was lighter than one made from steel and would not corrode. Unfortunately, the war broke out and automobile production gave way to churning out military equipment. The war also stimulated a great deal of research into making plastics from petroleum; these turned out to be more versatile than those derived from soybeans. Ford never did get back to the idea of a plastic car. Today, however, the idea of using soy products to make plastics has been resuscitated. In the 2002 model year, John Deere, one of the world's largest manufacturers of farm equipment, began using panels made of soybean and corn polymers on its combines. The new composite material is extremely strong and weighs 25 percent less than steel. The result, of course, is that these machines will consume less diesel fuel. This reduction is a big plus, since diesel releases copious amounts of pollutants. Unless the diesel we are talking about is biodiesel.

Scientists are certainly excited about the prospect of making fuel from a renewable agricultural source instead of from petroleum. German engineer Rudolf Diesel actually had this idea when he introduced his engine in 1900. His ideal fuel was peanut oil, but it turned out to be too viscous and gummed up the engine and had to be replaced by petroleum derivatives. Modern chemistry, though, has come to the rescue. Treating vegetable oils such as soybean oil with methanol and sodium hydroxide results in the production of biodiesel, which pollutes far less than regular diesel fuel. Even used restaurant grease can be turned into fuel through this process.

Joshua Tickell, an American energy consultant, dramatically demonstrated the potential of biofuel when he crisscrossed the US in his "Veggie Van." Along the way he collected used restaurant grease, which he converted to fuel using a miniature laboratory that he towed behind his vehicle. Tickell's undertaking was not pure theatrics. In France, diesel fuel already has to contain at least 5 percent biodiesel, and the US government has some vehicles running on 100 percent biodiesel made from soybeans. Henry Ford just might have been right when he predicted that "soybeans will make millions of dollars of added income for farmers and provide industry with materials to make things nobody even knows about now." But I doubt even Ford himself would have foreseen biodiesel fuel or aerodynamic cow chips.

FRENZIED MALES AND VIRGIN FEMALES

"Come quickly, Papa. Come and see these moths as big as birds!" cried the young child. And that is just what the nine-teenth-century French naturalist Jean-Henri Fabre saw as he rushed into his insect laboratory, where that morning he had incarcerated a freshly hatched female great peacock moth under a wire-gauze bell jar. Dozens of giant male suitors now fluttered through the open window and congregated around the prospective bride. Fabre wondered how they had known she was there, ready for a nuptial encounter. After numerous experiments with moths, he concluded that the males must be attracted by some sort of female scent. Although he was unable to come up with experimental evidence for his theory, Fabre's observation set the wheels in motion for the investigation of chemical communication between insects. The potential practical benefits of such research were realized by Joseph Lintner, an American entomologist who had amazed himself, as well as a crowd on the sidewalk below his office, by placing a female spice-bush silk moth on his window sill. Male moths from all directions quickly gathered to vie for the female's attention. "Cannot chemistry come to the aid of the entomologist in furnishing at moderate cost the odorous substance needed?" he asked, fully foreseeing the possibility of harnessing such chemicals to control insect pests.

It took almost a century, but Adolf Butenandt finally met the challenge. Butenandt was no ordinary chemist. At the age of thirty-six he had already won the Nobel Prize (1939) in chemistry for isolating the human sex hormones estrone and androsterone before turning his attentions to the insect world. Raising silkworms was a popular hobby in Germany at the time, and silkworm kits were common gifts for children. This wasn't surprising given that silk had captivated peoples' imaginations

since roughly 2700 B.C., when the Chinese princess His-Ling-Shi made an amazing discovery. As the story goes, the princess liked to take her tea under a mulberry tree in the imperial garden. One day a cocoon fell into the cup and when she tried to remove it, a fine thread began to unravel in her hand. The cocoon was acting as a virtual bobbin and had hundreds of feet of silk wound around it.

The Chinese realized that the silk thread could be woven into fine fabrics and recognized its commercial value; they managed to keep the origin of the alluring fiber a secret behind the Great Wall for close to 3,000 years! Eventually the world did learn that the key to silk is the silkworm—which isn't even really a worm. It is actually the caterpillar of the moth *Bombyx mori*. The female moth lays hundreds of eggs, which hatch into caterpillars (larvae) that eat nothing but mulberry leaves until they grow to be about 3 inches long. Then they begin to secrete silk fiber through their salivary glands and wind it around themselves to make a cocoon. Inside the cocoon the larva changes into a pupa, which in three weeks emerges as an adult moth. The adults do not eat and live only long enough to mate and lay eggs, starting the cycle over again. In commercial silk production the cocoons are dropped into hot water to kill the pupae, because the emergence of adults disrupts the silk threads. That's why some groups, such as the International Vegetarian Union, urge people not to buy silk. They claim the pupae are subjected to pain. One wonders if they are also opposed to swatting flies.

It was quite natural, then, for Butenandt, who knew about the pioneering work of Fabre and Lintner, to choose the silkworm moth for his chemical investigations. He knew about the frantic wing beating of the males in the presence of a female and recognized that this action was somehow related to the tiny gland that emerged from the tip of the female's abdomen at breeding time. Butenandt collected about 1 million cocoons from which he obtained 500,000 female moths. He proceeded to snip off the abdominal tips of these virgin females and grind them up. After years of work he managed to isolate 6.4 milligrams of a compound that turned out to be the female's sex attractant. "Bombykol," as the substance was named, thus became the first pheromone, or chemical communicator, ever isolated from an insect. Butenandt determined the exact molecular structure of this substance, a truly astounding achievement given the tiny amount of material he had to work with. The pheromonal activity was confirmed when bombykol was independently synthesized in the laboratory and shown to drive male silkworm moths into a frenzy.

And so began the field of insect pheromone research that has already paid huge dividends in controlling insect populations. The tomato pinworm, for example, tunnels into tomatoes and leaves pinholes and black blotches, destroying large portions of the crop. The female's sex attractant has now been isolated and synthetically reproduced. When released from little tubes attached to tomato plants, it interferes with the male moth's attempts to find a female. The prospective groom flies about following the false trails; he doesn't know if he's coming or going and dies a confused bachelor.

Silk research continues today. Why, those inventive Japanese have even come up with "spray-on" silk stockings. No more struggling to pull on panty hose. Just take a can of Air Stocking, a mix of finely powdered Plexiglas, silicone, and silk powder, and spray it on! I wonder what Cyd Charisse would think of this stuff. In 1957's *Silk Stockings,* the leggy actress played a stern communist sent to Paris to retrieve three wayward comrades who had succumbed to the evils of capitalism. Instead, her character, Ninotchka, warms to the charms of the West, symbolized by silk. In a classic scene, she slowly pulls silk stockings over her shapely legs as she gets ready to experience the capitalist charms of Fred Astaire. Somehow I doubt that seeing Charisse spray her legs with emulsified silk powder would hold the same allure.

MUNCHING ON A CHEMIST

You know that I love chemists. And right now I'm going to eat one. He's fully cooked, but I expect he would probably taste pretty bland without the added natural and artificial flavors. For my dining pleasure he has been encased in hardened maltitol syrup and colored with Yellow Dye #5 and Blue Dye #1. Rigor mortis has now set in, but in his heyday this guy was pretty adept at carrying out some amazing chemical reactions. Like converting plant extracts into aphrodisiacs.

If you're wondering, no, I haven't taken leave of my senses. Actually, I'm using them. I'm sitting here sucking on "Worm

Candy," which looks like a regular lollipop except for the clearly visible "worm" that is smack in the middle. In reality, it is a 1-inch-long insect larva that the candymaker, exercising some literary license for shock value, has called a worm. And why am I doing this? Because it allows me to call attention to the remarkable chemistry of insects.

Let's start with the salt marsh moth. This is not the insect that dines on the woollies in your closet. The larvae of this moth prefer plants, especially those that contain naturally occurring pyrrolizidine alkaloids. Plants, of course, are veritable chemical factories and convert carbon dioxide, water, and nutrients from the soil into thousands of compounds. All the proteins, fats, carbohydrates, and vitamins we need to sustain life originate in plants, which we eat either directly or through animal intermediaries. Plants, however, are not always keen to sacrifice themselves as food for others and have evolved various protective mechanisms. Insects are among the most notorious plant predators, and it therefore comes as no surprise that many plants produce a variety of natural insecticides. Have you ever wondered, for example, why tobacco plants produce nicotine, or why the coca plant synthesizes cocaine? Both chemicals deter insects. Nicotine is actually used as a commercial insecticide, and researchers have shown that cocaine may be an even more effective one. They sprayed cocaine solution on the leaves of tomato plants and then placed moth caterpillars on these leaves. The insects reared up, began to shake, and turned away from the cocaine, apparently showing greater intelligence than some humans.

Pyrrolizidine alkaloids are a class of natural toxins that plants use to wage chemical warfare against insects. With a few exceptions, they are not found in plants normally consumed by humans. That's a good thing, because in a sufficiently high dose these compounds are toxic to the liver. Comfrey is one of the plants that contains pyrrolizidines and, curiously, is sometimes recommended by herbalists in the form of a tea for improved health. Admittedly, it takes large amounts of comfrey to cause liver toxicity, but it has happened. So how is it, then, that the caterpillar of the salt marsh moth happily frolics on the leaves of pyrrolizidine alkaloid–producing plants and even makes a meal of them? The development of resistance to toxins is quite common in the insect world, as farmers who routinely face this problem with commercial insecticides well know. It seems that salt marsh moths not only have developed a resistance to pyrrolizidine alkaloids but also have turned them into an evolutionary advantage. They convert the toxins to a compound called "hydroxydanaidal," which has some truly fascinating effects and may well deserve to be called a "moth aphrodisiac."

The male salt marsh moth has little inflatable organs on its belly called "coremata." These tube-like appendages are covered with scent scales, which protrude like tiny hairs. Their purpose is to provide a large surface area through which the scent of hydroxydanaidal can be wafted into the air. When the female senses this chemical, she comes a-running and mating ensues. The more effective this chemical release is, the greater the chance that the suitor will attract a mate. Amazingly, the male can increase his chances of romance simply by eating right. When scientists put salt marsh moth larvae on diets with different amounts of pyrrolizidine alkaloids (believe it or not, "salt marsh caterpillar diet" is commercially available from Bioserv Inc. of Frenchtown, New Jersey), they were able to show that adult males fed the largest amounts of the hydroxydanaidal precursors developed the largest coremata. When fully erect, the coremata of the high-dose males reached an impressive 2 centimeters whereas the pyrrolizidine-deprived moths barely managed to muster up 0.5 centimeters. Just wait till those spam e-mailers get hold of this information: "Not satisfied with the

size of your coremata? We've got the answer! Pyrrolizidine alkaloids—all natural, clinically proven!"

Now let's turn to the scarlet-bodied wasp moth. The male doesn't use alkaloids to attract the female, but he does offer up a nuptial present of these chemicals that may save her life! Dog fennel is a plant eschewed by herbivores because of its pyrolizidine content, which makes it very bitter. But the scarlet-bodied wasp moth just loves it. He ingests the plant's juices and stores some in tiny pouches under his abdomen. When the moth engages in amorous activities with the female, he transfers some of the juice to her, instantly making her taste awful to predators. When researchers mated virgin females both with males that were bred on the alkaloids and with males that were not, and then left the unfortunate moths at the mercy of a notorious moth predator, the golden silk spider, they found that the unprotected insects were quickly eaten. The noxious males and their chemically protected concubines escaped.

Frankly, I don't know what sort of caterpillar is in my lollipop. The ingredients simply identify it as "insect larva." You see what happens when you have poor labeling laws? But I've now consumed most of the candy and am almost down to the "worm." OK, here we go. Crunch! Not bitter at all. Not a hint of pyrrolizidine alkaloids. Kind of tasty, actually. Best little chemist I've ever eaten. High in protein, low in fat. You should try one.

ONCE UPON A TIME

IT STARTED WITH A BANG

This story starts with a bang. A bang heard by young Justus Liebig sometime around 1810 in the marketplace of the small German town of Darmstadt, where an itinerant showman was entertaining the crowd with a series of chemical demonstrations. When some overeager boys approached too closely, he scattered them with little packets of explosive chemicals that he pitched at their feet. Liebig was impressed and managed to find out that these "torpedoes," as the little packages were called, were made from mercury, nitric acid, and alcohol. Since his father owned a shop that dealt in chemicals, procuring these reagents would not be a problem. Indeed, his father would probably not object to the chemical dabbling, since the elder Liebig was himself a keen experimenter and known locally as a "sorcerer." But it would be his son, the sorcerer's apprentice as it were, who would achieve lasting fame with his chemical experiments.

Young Liebig did manage to produce some torpedoes, which he then sold in his father's store. Making the required mercury "fulminate" ($Hg[ONC]_2$), though, was not an easy task. After a batch exploded prematurely, Liebig Senior decided that perhaps his son's days would be better spent in the local school. Philosophy, history, and mathematics were, however, not to the boy's liking. When the school's headmaster sarcastically asked him what he thought he was going to be, Liebig replied, "A chemist." This answer elicited derisive laughter, as no such profession existed at the time. In the end, Liebig would have the last laugh.

Seeing that his son was unhappy at school, Justus's father apprenticed him to an apothecary. As Liebig described in his later lectures, this was not an altogether happy arrangement and ended with yet another laboratory bang. At this point, the elder Liebig didn't know what to do with his son. But one of his customers, chemistry professor Karl Kastner, did. Despite the youngster's little formal education, he took the boy under his wing and arranged to enroll him at the University of Bonn. Liebig turned out to be a brilliant student. So brilliant, in fact, that after a short tenure at the Sorbonne in Paris, he was offered a position as "extraordinary professor" of chemistry at the University of Giessen in Germany at the remarkably young age of twenty-one. There he quickly established what was to become the most famous teaching laboratory in Europe, drawing students from all over the world. They mostly came to learn about Liebig's techniques in analytical chemistry. The professor had decided early on that the key to chemical advance was the ability to determine the exact composition of substances. He developed a method of combusting organic compounds and capturing the carbon dioxide and water that formed in a solution of caustic potash (potassium hydroxide). From the weight of potassium carbonate generated, he was able to calculate the amount of carbon and hydrogen in the original sample. Students still carry out analyses using the Liebig Apparatus today.

Liebig never gave up his love of the explosive fulminates. Indeed, his continued work on these substances led to a breakthrough in chemical thinking. He became aware that subjecting silver fulminate to chemical analysis yielded the same results as an analysis of silver cyanate that had been synthesized by his good friend and noted chemist Friedrich Wohler. The cyanate did not have the same explosive properties as the fulminate. How could two substances have the same chemical composition and different properties? The only logical possibility was that the same atoms were somehow joined together in a different fashion. Indeed, silver cyanate turned out to be AgCNO, and silver fulminate AgONC. With this observation Liebig introduced the idea of "isomers," compounds composed of the same elements but bonded together in different ways.

Around 1840, Liebig began to take an interest in the chemistry of living things. This curiosity marked the beginning of the second phase of his career, a phase that would be far more controversial than his early work on chemical analysis. He steadfastly denied, in spite of the evidence, that yeasts were living organisms. Stubbornly, Liebig insisted that sugars turned into alcohol when putrefying yeast cells communicated some sort of molecular vibration to the solution. When he discovered nitrogen compounds in urine, he postulated that they must stem from the breakdown of muscle during activity and concluded that extra proteins (which are the body's prime source of nitrogen compounds) should be supplied in the diets of athletes. This theory gave rise to the "steak before a game" idea and even led Liebig into the commercial production of Liebig's Fleisch Extract for "benefits to the body." When the Oxford rowing team adopted the Liebig formula and routinely defeated archrival Cambridge, the benefits of the extract seemed established. It didn't appear to matter to anyone that the Oxford crew actually trained much harder than its rivals. Competing protein extracts soon appeared, each proclaiming its restorative and invigorating powers. One of these extracts, Bovril, is still with us, although after some pioneering research in Germany in the latter part of the nineteenth century, it was no longer touted as a performance-enhancing substance.

Max von Pettenkofer and Karl von Voit simply measured nitrogen output in the urine of athletes and showed clearly that protein was not being consumed during vigorous exercise. Paradoxically, Liebig had never carried out such an experiment in spite of his public exhortations about the importance of chemical analysis. He did, however, make use of his analytical talents to determine the mineral composition of plants and established the basis for the modern fertilizer industry, although here too he made some fundamental errors. Liebig believed that nitrogen could be supplied to plants from the air and arrogantly belittled the use of nitrates. He also failed to understand the use of manure to give structure to the soil and overemphasized the use of mineral salts.

In spite of these foibles, Justus von Liebig (he was made a baron in 1845) was one of the greatest chemists of the nineteenth century. Still, that didn't stop Nazi vandals, who believed that Liebig's mother had been Jewish, from splattering his monument in Munich with potassium permanganate and silver nitrate just prior to World War II. Later, authorities had a hard time cleaning the statue. Herr Professor Liebig was no longer around to give them what surely would have been good chemical advice.

TAKING THE FUZZY OUT OF FIZZ

In 1767 Joseph Priestly was to take up his position as minister at Mill Hill Chapel in Leeds, England. But when he arrived, he discovered that the minister's house, located next to the chapel, was not ready and his family would have to make do with a temporary residence, which was next to a brewery. Had it not been for this quirk of fate, the progress of chemistry could have taken a completely different route.

Priestley could not help but notice the vapors given off by the brewery. He became very interested in these "airs," as he called them, particularly in the one that was responsible for the bubbles in beer. This "fixed air," as carbon dioxide was known at the time, seemed to be the same gas that made certain naturally occurring spring waters effervescent. Health resorts in Europe were serving such fizzy waters as supposed cures for various illnesses, and Priestley began to wonder if an artificial fizz could be added to ordinary water.

Joseph Black had already shown that "fixed air" could be produced by the action of acids on marble. So Priestley combined sulfuric acid and calcium carbonate to form carbon dioxide—although he did not, of course, recognize the gas as such. He collected the gas in a pig's bladder and found a way to use it to carbonate water. He was awarded the Royal Society's prestigious Copley Medal for his publication *Directions for Impregnating Water with Fixed Air.*

"Soda water," as the fizzy stuff was called, became very popular. It was taken along on ocean voyages because it tasted better than the usual stored, stagnant water. It also developed a false reputation as a preventative against scurvy and other diseases and was actually sold in apothecary shops. But John Nooth, a Scottish physician, complained that the use of a pig bladder imparted an off flavor to the water. He developed a glass apparatus for carbonating water to solve the problem. This method found widespread use in shops and homes. The soda boom had begun.

His success at carbonating water had whetted Priestley's appetite for science. A poor minister, however, could hardly afford to devote his time to the investigation of "airs." He needed a rich patron. William, the second Earl of Shelburne, had heard about Priestley's experiments and invited him to come to Bowood, about 90 miles west of London, to carry out scientific investigations under his patronage. And so it happened that on a sunny day in August 1774, perhaps the single most important experiment in the history of chemistry was performed in a laboratory on the estate of the Earl of Shelburne. Using a magnifying glass, Joseph Priestley focused the sun's rays on a sample of red calx (mercuric oxide) and noted that an "air" was given off and that it was insoluble in water. A candle burned in a spectacular fashion when exposed to the gas and a mouse became more vigorous when confined to a jar filled with it. Eventually Priestley himself inhaled the gas and remarked, "Who can tell but that, in time, this pure air may become fashionable article in luxury. Hitherto only two mice and myself have had the privilege of breathing it."

Little did Priestley realize that he, the mice, and every living animal had been inhaling his newly discovered gas with every breath. Joseph Priestley had isolated pure oxygen! While Priestley was a great experimentalist and observer, he was not an astute interpreter of his experiments. He never recognized oxygen for what it was. Priestley firmly believed that what he had created was "dephlogisticated air." At the time the prevailing opinion was that substances that burned did so because they contained "phlogiston," which was released into the air during combustion. When the air became saturated with phlogiston, it would no longer support combustion. That's why a candle burning inside a closed jar would be extinguished. So it made sense to Priestley that his "dephlogisticated air" would be able to take up more phlogiston and that his candle would burn longer and more brightly.

Soon after his classic experiment Priestley accompanied the earl on a trip to the Continent where he met Antoine Lavoisier, the noted French scientist. He carefully described his mercuric oxide experiment to Lavoisier, who not only repeated it but also interpreted it correctly. Lavoisier identified oxygen as an element and determined that air was composed of two substances, one of which supported combustion and one that did not. The latter he named *azote,* from the Greek for "no life," which is still the French term for nitrogen. Interestingly, the Swedish chemist Carl Wilhelm Scheele had independently isolated oxygen at least a year before Priestley's discovery but did not publish his work until 1777. Priestley had reported his results immediately. It pays to publish!

For some unknown reason, Priestley had a falling out with the Earl of Shelburne and returned to his first calling, being a Unitarian minister, and moved to Birmingham. This relocation did not have a happy outcome. Priestley was a supporter of the principles of the French and American revolutions and, as a result, a royalist mob ransacked his house and destroyed his laboratory in 1791. Priestley thought America might be more tolerant, so he pulled up his roots and moved to Northumberland in Pennsylvania where he continued his experiments, eventually discovering and isolating carbon monoxide. Both Priestley's house in Pennsylvania and Bowood mansion in England have been designated as historic landmarks by the American Chemical Society, which itself was first conceived at Joseph Priestley's gravesite.

It was there that in 1874 thirty-five men gathered to commemorate the centennial of the discovery of oxygen and contemplated the notion of a national chemical society. Two years later the idea crystallized into the formation of the American Chemical Society, the largest professional organization in the world.

One wonders what Priestley would think about the currently popular oxygenated-water scams. "Purifies your bloodstream and eliminates toxins," scream the ads. Given that the solubility of oxygen is about 7.5 parts per million, a dose of this miraculous water actually contains less oxygen than a single breath. Assuming that the oxygen from the beverage actually gets into the bloodstream, roughly 3 liters of water per minute would have to be consumed to increase blood oxygen by 1 percent. This silliness reminds me of another of Priestley's discoveries: laughing gas.

TAKE A BROMIDE!

What is common to Florence, Italy, and Baltimore, Maryland? Not much, one would think. But if you walk along West Lombard Street in Baltimore and stop at the corner of South Eutaw Street and gaze up, you just might think you're in Florence, looking at the tower of the Palazzo Vecchio. Except for one thing. The top of the Emerson Tower in Baltimore is bathed in blue light. Why blue? Because that was the color of the bottle Bromo-Seltzer originally came in. The replica tower was commissioned by Isaac Emerson, the inventor of the famous product that began to cure Americans' headaches about 100 years ago. The tower was originally adorned with a huge Bromo-Seltzer bottle, which had to be removed in 1936 for safety reasons and has now been replaced with a beautiful blue glow. Emerson graduated with a degree in chemistry from the University of North Carolina and, with financial backing from his new wife, opened up three small drugstores in Baltimore. It was behind the counter of one of these that he developed his classic formula.

The late 1800s were the halcyon days of the patent medicine era, and numerous pain-relieving nostrums clamored for the public's attention. Emerson realized that if he was going to be successful, he needed something to distinguish his product from the others. The answer occurred to him thanks to his background in chemistry. When dissolved in water, sodium bicarbonate (baking soda) and citric acid combine to produce bubbles of carbon dioxide. "Fizz" would certainly make his product different. But it wouldn't cure a headache. At the time, acetanilide was already a widely sold pain reliever, so Emerson decided to add it. Still, many of his competitors' products had this ingredient. He needed something else. Some have suggested an idea came to him when he read about the eruption of Mt. Bromo, a volcano on the island of Java. Whether spurred by that event, or just by his knowledge of chemistry, he decided to add some sodium bromide to the concoction as well. Emerson knew that bromides had sedative properties and could therefore be useful in the treatment of "tension headaches." Indeed, bromides had entered medical practice in 1857 when Charles Locock, a physician from London, England, described how he had used potassium bromide to treat a patient with "hysterical epilepsy." He went on to say that he had even tried it "in cases of hysteria in young women, unaccompanied by epilepsy," finding it "of the greatest service." In Victorian times, "hysteria" was a common diagnosis in young women; its cause was thought to be some disturbance in the womb (the term itself derives from the Greek *hysterikos,* meaning "of the womb"). It didn't take long for potassium bromide to be established as a treatment for various nervous conditions.

Bromides actually do have sedative properties, so their use was not nonsensical. Their mechanism of action is thought to be the close resemblance of the bromide ion to chloride. Chlorides are important in the functioning of nerve cells, and chloride overload can cause overstimulation. Since bromides mitigate chloride activity, they can act to reduce seizures and to sedate. Until the advent of the barbiturates, potassium bromide was widely used and even gave rise to the expression "Take a bromide," meaning "Calm down."

So sodium bromide would be Emerson's other ingredient. But there was yet another component necessary to the Bromo-Seltzer mixture, perhaps the most important one: ingenious advertising. First, Emerson wanted his product to have a really unique look, so he packaged it in a glorious cobalt-blue bottle. Then, he decided, it would be promoted as a medication that fights a headache in three ways. As Americans were to learn, especially fans of the radio program *The Adventures of Ellery Queen* sponsored by Bromo-Seltzer, there was more to a headache than just a pain in the head. An upset stomach and jumpy nerves also played key roles in the misery. A "sick headache" was the result of this triple whammy. Bromo-Seltzer contained acetanilide to ease the pain, bicarbonate to relieve excess stomach acidity, and sodium bromide to calm jumpy nerves. And there was another benefit: Bromo-Seltzer dissolved immediately in water so it was ready to go to work right away to relieve the headache, settle the stomach, and soothe the nerves. The noisy fizz, of course, also helped to convince the patient that this was a really active medication. Not everyone, though, liked the loud fizz. A waiter once offered a hung over W. C. Fields some Bromo-Seltzer. "No-o-o-o," the comedian moaned, "I couldn't stand the noise."

Most people could stand the noise all right; it was the other ingredients that caused them problems. Problems like their skin turning blue! This mystery was solved by Bernard Brodie and Julius Axelrod (a Nobel Prize winner in 1970), who discovered that acetanilide, the painkiller in Bromo-Seltzer, could cause methemoglobinemia, a condition in which hemoglobin loses its ability to bind oxygen. Since oxygenated blood is red and blood that lacks oxygen is bluish, people who suffer from methemoglobinemia develop a blue tinge. Brodie and Axelrod suggested that acetanilide in Bromo-Seltzer be replaced with acetaminophen, the pain reliever we are now familiar with as the active ingredient in Tylenol. The manufacturer complied and the blue problem vanished. But there was yet another difficulty.

Some customers who used Bromo-Seltzer exhibited symptoms of bromine toxicity, known as "bromism." Symptoms range from sluggishness and slurred speech to confusion and acne-like eruptions on the skin. This response eventually led to the removal of sodium bromide from the product, leaving the modern version with just acetaminophen and fizz. Bromo-Seltzer is no longer capable of causing bromism, but that doesn't mean the condition has disappeared. Recently, a patient showed up at a medical center in Cleveland, Ohio, with postules on the hands that smacked of bromism. But it took a while for the physicians to figure out what was going on. It seems the cause was the consumption of brominated vegetable oil, which is a solvent used to dissolve flavors that are to be added to citrus beverages. The culprit turned out to be Ruby Red Squirt, a citrus-flavored soda pop. But normal people don't have to be concerned about developing bromism from this product. The man in question was drinking—get this—8 liters of the stuff every day! And to those who try to use this account to highlight the dangers of food additives, I have only one thing to say: Take a bromide.

Calm Down

"Case 1: W. B., a male, aged fifty-one years, who had been in a state of chronic mania excitement for five years, restless, dirty, destructive, mischievous and interfering, had long been regarded as the most troublesome patient in the ward. His response was highly gratifying. From the start of treatment with lithium citrate he steadily settled down and in three weeks was enjoying the unaccustomed surroundings of the convalescent ward." So began the report in which Dr. John Cade described how he had successfully treated ten manic patients. The paper, entitled "Lithium salts in the treatment of psychotic excitement," was published in the September 3, 1949, issue of the *Medical Journal of Australia*. The journal's readership at the time was very limited, but the article's eventual impact was gigantic. John Cade had alerted the medical world to the medication that would eventually benefit millions of manic patients.

Mania, in a sense, is the opposite of depression. A manic patient is in a state of elation, is often overtalkative, doesn't sleep much, and has flighty ideas and poor judgment. Shopping sprees and sexual promiscuity are also characteristic of mania. While some patients may be affected only by mania, it is more common for people to experience periods of mania alternately with periods of depression—hence the description of these individuals as "manic-depressive." Often the cycle of mania and depression is so regular that family members can predict the future state of mind of a manic-depressive patient. Vacations, for example, can be planned around manic, instead of depressive, periods. The incidence of the disease is greater among creative people. Mark Twain, Cole Porter, and Vincent van Gogh are interesting examples of people who were supposedly affected by "bipolar disease," as the condition is also known. The illness has a genetic component; if one identical twin has the disease, chances are greater than 70 percent that the other will also show symptoms.

The story of John Cade's discovery of lithium therapy begins in a prisoner-of-war camp in Singapore where he spent three years during World War II. Here he kept track of the mental state of his fellow prisoners and noted that those who exhibited signs of mental illness did not do so regularly. They had good days and bad days; periods of euphoria alternated with periods of depression. This alternation seemed to smack not of a permanent physical injury to the brain, but rather of some sort of chemical imbalance.

Cade hypothesized that mental illness was caused by a toxin that entered the brain and was subsequently excreted in the urine. As the toxin was eliminated, the symptoms vanished. What could the toxin be? After the war, Dr. Cade began a systematic search by injecting the concentrated urine from manic patients, schizophrenics, and healthy people into guinea pigs. He made the rather unstartling observation that guinea pigs injected with human urine died. This was no earth-shaking discovery, but the psychiatrist apparently noted that the animals injected with the urine of manic patients died more quickly.

Dr. Cade suspected that the poison in the urine was urea; when he administered this substance to the guinea pigs, the animals died the same way they had when injected with the urine of manic patients. But when he tested the urine of manic patients for urea, he did not find that levels were higher than in other urine samples. Perhaps there was something else in the urine that modified the toxic effects of urea. His attention turned to uric acid, which at the time was probably the best-studied urine component. Doctors already knew, for example, that high levels of uric acid caused the terrible pains of gout. Cade thought that maybe high levels of uric acid also made urea more toxic.

He decided to inject urea and uric acid together into guinea pigs to study the effect. There was a problem, though: uric acid was not very soluble in water. Remembering his chemistry, Dr. Cade knew that acids could be converted to soluble salts by neutralizing them with a base. Fortunately (as it later turned out), the base he chose was lithium carbonate, which formed soluble lithium urate. Much to his surprise, the lithium urate actually reduced the toxicity of urea! The next logical thing to do was to co-inject lithium and urea to see if the protection was due to lithium. It was! Now, still thinking that urea may somehow be responsible for mania, Dr. Cade wondered about using lithium carbonate as a treatment. First, he had to test its safety in guinea pigs. When he injected them with the drug, he noted that they became remarkably calm. Could it also calm his mentally ill patients? He decided to find out.

Dr. Cade treated ten manic patients—six schizophrenics and three who suffered from mania and depression—with lithium carbonate. All the manic patients improved, some dramatically. Given that no other treatment for mania was available at the time, one would think that news of this success would have spread like wildfire. It didn't. The *Medical Journal of Australia* was not commonly read by most physicians. Fortunately, however, Mogens Schou, a young Danish psychiatrist, did read it and was intrigued enough to further the exploration of lithium therapy. It was his work that was largely responsible for the introduction of lithium carbonate as a recognized treatment for manic-depressive illness in 1970.

Is lithium a wonder drug? Of course not. All drugs come with some baggage. Possible lithium side effects include trembling hands, lethargy, nausea, weight gain, and frequent urination. Furthermore, about 30 to 40 percent of patients do not respond to lithium. But there is no doubt that John Cade's rather fortuitous discovery has eased the life of numerous manic-depressive patients.

There is a curious footnote to this account. In 1929 Lithiated Lemon-Lime Soda was introduced with the slogan "Takes the 'ouch' out of 'grouch.'" It was a huge success. Until the 1940s, the beverage listed lithium on its label as an ingredient. Maybe it really did have an effect on consumers' mental states. What ever happened to Lithiated Lemon-Lime Soda? Its name was changed to 7-UP.

A Poison Eater Who Just Said, "No"

I would have loved to have been in the audience at the fabled Argyll Rooms in London, England, on February 4, 1829. The theater was packed with men and women who had come to witness an epic confrontation between Monsieur Ivan Chabert, the Fire King and Poison Eater, and Dr. Thomas Wakely, the editor of the prime British medical journal *The Lancet*. Chabert was the talk of the town, especially after he had won a wager from a Mr. J. Smith, who had accused him of trickery and claimed to be able to reproduce all of the Frenchman's feats.

Those feats were truly amazing. Chabert would walk into a specially built oven with raw steaks in his hands and emerge unscathed with the meat fully cooked. He would then proceed to bathe his feet in boiling lead, rub a red-hot shovel on his tongue, inhale arsenic vapors from a fire, and finish the proceedings with a meal of poisonous phosphorus. Pretty entertaining stuff! I think it beats today's "reality" programs, which feature contestants downing nothing more dangerous than pig snouts or cockroaches. Mr. Smith, though, wasn't amused and didn't believe that Chabert's effects were legitimate. He launched a challenge, putting his money where his mouth was. Which is exactly where Chabert put the phosphorus. It seems that the entertainer really did eat 20 grams of phosphorus to the satisfaction of Smith, who declined a similar meal and paid up. This settlement triggered allegations of chicanery as well as collusion for purposes of publicity.

Accounts of the Smith episode are too sketchy to allow conclusions to be drawn about what actually transpired, but Dr. Wakely's challenge at the Argyll Rooms is well documented. Wakely was not too interested in how Chabert kept from cooking himself in the oven; he assumed that it was constructed in a way that allowed some cooling ventilation in the spot where the "Fire King" baked while his steaks roasted. But the phosphorus business really intrigued him. As far as this stunt was concerned, Wakely thought there were only two possibilities: Chabert either used sleight of hand and never swallowed the phosphorus or underwent a quick gastric lavage. Wakely suspected the latter, because Chabert usually retired from the stage after the phosphorus meal, supposedly to change into appropriate clothes for the next part of his repertoire, which was the oven performance. But what really got Wakely's goat was Chabert's claim that he could swallow what would be a lethal dose of cyanide, thanks to his discovery of a marvelous antidote.

Dr. Wakely challenged Chabert to take cyanide in public and in his presence. The entertainer was forced to accept the test; after all, he was being asked to do no more than what his playbills loudly proclaimed. Being a physician, Wakely had some ethical concerns about the proposed proceedings. He thought he might be held accountable for Chabert's death, which he believed would be a certainty if the man really were foolish enough to take the cyanide. Convinced that Chabert would balk, however, he pressed ahead. And balk Chabert did. The "Poison Eater" told the crowd that he had never claimed he would take the cyanide himself; he was only going to demonstrate the properties of the antidote in a dog. He was prepared to carry out his demonstration, he said, but he never got the chance. Amid cries of "faker" and "cheat," the angry crowd jostled him out of the theater, and Chabert had to take refuge in a neighboring cellar.

Surprisingly, this debacle did not put an end to Chabert's career. He continued with his performances, on one occasion even resuscitating his cyanide antidote claim and putting it to a test. First, an unfortunate dog was given a dose of cyanide and, as expected, died within a few minutes. Then a second animal was then brought out and fed the cyanide and the antidote in quick succession. In fact, the feeding was so quick that some of the onlookers claimed administering the antidote was just a subterfuge for removing the cyanide from the animal's mouth. The dog survived, but the experiment was never repeated and Chabert never again mentioned cyanide. Interest in this brand of flimflammery faded in England and Chabert decided to seek greener pastures across the ocean.

In New York, billing himself as a "Professor of Chemistry and Pyrotechnic Arts," Chabert tried unsuccessfully to replicate his European success. Then he realized that burning the public was far more profitable than burning steaks. Chabert awarded himself a medical degree and, as Xavier Chabert, MD, began to market products such as Tapuyas Elixir for the treatment of toothaches and scurvy. He also claimed to have invented a remedy for cholera, which firmly implanted him in the realm of the quacks and eventually brought him to the attention of that great scourge of charlatans, Harry Houdini. Indeed, it was the magician who tracked down the historical records that offer us a glimpse into the career of the Fire King and presented them in his wonderful book *The Miracle Mongers: An Exposé*.

While Houdini took careful aim at performers he felt hoodwinked the public, he praised those who entertained with scientific legerdemain. An example is Floram Marchand, one of a number of "water conjurers" who, from the seventeenth century on, amazed the public by spouting a variety of beverages from their mouths. Marchand would drink water and out would come beer, then wine, then claret! The secret was to first swallow an extract of Brazil wood, which contains brazilin, a natural indicator that is yellow in acid and red in base. The acid contents of the stomach ensured that yellow "beer" would be regurgitated when a large amount of water was swallowed. Then, if the contents were spouted into a glass secretly prepared with some alkali at the bottom, "wine" would be produced. Drinking more water diluted the dye and yielded claret. The water con-jurors never claimed to have supernatural powers, so skeptics swallowed their acts more readily than those put on by im-posters such as Chabert. I would have liked to have seen Marchand's water spouting. I think it probably would have been a lot more entertaining than watching the Osbournes spout nonsense.

You're Feeling Sle-e-e-epy

The eagles seemed to be falling out of the sky near Kodiak, Alaska. Those that survived staggered around in a stupor, as if they were drunk. Undigested meat was found in the dead birds' stomachs, suggesting that they had succumbed soon after eating. Analysis of this meat revealed traces of pentobarbital, a com-monly used sedative. What meat had the eagles eaten and how had it become contaminated with the barbiturate? Since each living organism has a characteristic protein pattern, a careful analysis of the meat proteins was carried out. Results for deer, sheep, rodents, and bear—all animals the eagles could have dined on—came up negative. But a positive match was found for cats! Nobody, though, had reported cats being pilfered by eagles, so what was going on? The key to the mystery was the location of the birds' roosting site. It was near the town's landfill! And right there, on top of the garbage, lay several cats that had been euthanized by the local animal shelter. Instead of burying or incinerating the animals, their bodies had been dumped into the garbage. It was the residues of pentobarbital, a frequently used euthanizing agent, which had so dramatically affected the eagles.

Barbiturates are very effective sedatives and sleep inducers. When you see a veterinarian subdue an animal with a tranquil-izer dart in a wildlife documentary, chances are you're seeing a fast-acting barbiturate in action. These same drugs are commonly used by anesthetists. Injecting certain barbiturates intravenously puts a patient to sleep almost instantly. The effect does not last very long, but certainly long enough to induce the deep hypnotic state needed in order to administer surgery. As with any drug, dosage is critical. Too high a dose of barbiturates can induce permanent sleep. Marilyn Monroe's death is perhaps the most widely publicized example of barbiturate overdose, but her case is certainly not unique; barbiturates are probably the drugs most commonly used to commit suicide. Sometimes these "suicides" may be inadvertent. Both barbiturates and alcohol are central nervous system depressants, and their synergistic effect can be lethal. Many a victim has been found with a bottle of barbiturates and a glass of some alcoholic beverage sitting on the night table.

Interestingly, the first barbiturate ever made didn't have any pharmacological activity. Way back in 1864, German chemist Adolph von Baeyer combined malic acid isolated from apples with urea from urine to make barbituric acid. This compound is actually pretty uninteresting except, perhaps, for its name, which to this day provokes controversy. Some claim that its origin lies in the Latin *barba,* for "beard," since at the time chemists would supposedly shake their beards over a solution that refused to crystallize. Bits of dandruff or perhaps crystals from previous experiments (hygiene may not have been optimal in those days) acted as "seeds" around which crystals could form. Another theory is that a waitress named Barbara had provided the urine sample that was used to isolate the required urea. A thrilling account to be sure, but probably not factual. The most reasonable story is that on the day von Baeyer first synthesized barbituric acid, he had visited a local tavern where soldiers were hoisting a few pints in honor of St. Barbara, the patron saint of artillerists and miners.

As is generally the case when a new compound is synthesized, chemists begin to tinker with its molecular structure, hoping to find a useful derivative. It took some forty years of such chemical dabbling before research on barbituric acid derivatives paid off. But when it did, it paid off big. In 1903 Emil Fischer, a former student of von Baeyer's, synthesized diethylbarbituric acid, and Joseph von Mering, his physician collaborator, showed that it readily put dogs to sleep. Von Mering proposed the name "Veronal" for the new drug after the Italian city of Verona, which he considered to be the most peaceful place he had ever visited.

Why was Veronal active while barbituric acid had no effect? Because the two "ethyl" groupings (2-carbon fragments) that had been added to the molecule increased its fat solubility and facilitated its passage into the bloodstream through the tissues of the intestine. Actually, as far as inducing sleep went, Veronal worked a little too well. The effect was slow to wear off and people felt sleepy the day after taking it. This drawback precipitated the synthesis of hundreds of derivatives of barbituric acid in an attempt to find more effective drugs. An "improved version," phenobarbital (Luminal), followed in 1912 and was later joined by the likes of pentobarbital (Nembutal), secobarbital (Seconal), and amobarbital (Amytal). All have been useful in the treatment of anxiety disorders, tension, panic attacks, and sleep problems. Unfortunately, they can also be habit forming and have the potential to become drugs of abuse.

Since barbiturates so readily induced sleep, they caught the attention of anesthetists. In the 1930s, Ernest Volwiler and Donalee Tabern at Abbott Laboratories, a pharmaceutical company, sought a substance that could be injected directly into the bloodstream to produce rapid unconsciousness. They labored for three years and screened over 200 compounds. Simply replacing an oxygen atom with sulfur in pentobarbital finally ended their quest. Sodium pentothal turned out to be an ideal agent for putting patients to sleep quickly. And it had another interesting property. Just before losing consciousness, some patients began to babble in the most uninhibited fashion, revealing their innermost secrets. Could sodium pentothal serve as a "truth serum"? Authors and moviemakers were quick to weave it into their plots, but the scientific community is divided on the reliability and, of course, the legality of information extracted with sodium pentothal. I would guess that the CIA and the former KGB may have some insight into these matters.

Although Emil Fischer made numerous contributions to chemistry (he received the 1902 Nobel Prize), his name, at least in the public's mind, was most readily associated with Veronal. Indeed, novelist Hermann Sudermann complimented Fischer on his discovery: "You know, it is so efficient, I don't even have to take it; it's enough that I see it on my nightstand." Fischer's ingenuity was apparently not restricted to synthesizing sleeping pills. "What a coincidence," he replied. "When I have problems falling asleep, I take one of your novels. As a matter of fact, it's enough that I see one of your wonderful books on my nightstand and I immediately fall asleep!"

MOLECULES AND MIRRORS

It was a good idea both scientifically and economically. The patent protection on pharmaceutical giant Eli Lilly's blockbuster antidepressant Prozac was coming to an end and competitors were getting ready to market generic versions of the drug. Lilly obviously did not relish the generics taking a bite out of Prozac's annual $2.5 billion in sales and vigorously explored ways in which the patent protection could be extended. A new version of Prozac would do the trick. And it could be done, in a sense, with mirrors!

To understand this plan we need to leave behind the impressive, well-equipped research laboratories of the twenty-first century and travel back in time to 1848 and a dimly lit laboratory at the École Normale Supérieure in Paris, France, where a young student, the son of a tanner, was toiling away on his doctorate in chemistry. Louis Pasteur was studying the sediment formed during the fermentation of wine. Two substances had previously been isolated from such sediments, with the main component christened "tartaric acid" and the minor one named "racemic acid" after the Latin expression for a "bunch of grapes." These two acids had the same chemical composition and were virtually identical except for one subtle difference: they behaved differently when placed in a beam of "plane-polarized light."

Light is a most unusual phenomenon and is difficult to discuss in a simplistic fashion. But perhaps we can take a shot at it by thinking about 3-D movies. The effect here relies on one eye seeing a slightly different image than the other. The two images can clearly be seen when the screen is viewed with the naked eye, but they merge into one three-dimensional scene when viewed with special glasses. The glasses are constructed in such a way that each lens allows only the light from one of the images on the screen to pass through. It may help to think of light as being composed of waves vibrating in every direction and the filter as being a picket fence that allows only those rays of light that vibrate in the same plane as the spaces between the slats in the fence to pass through. The two lenses in the glasses can then be thought of as picket fences that are perpendicular to each other; the two images on the screen are projected through two similar lenses. Each eye therefore sees a slightly different scene and the brain meshes them together into a three-dimensional image.

A simple experiment can help to make sense of this unusual phenomenon. Take a pair of 3-D glasses and break them in half. (A cheap pair of Polaroid sunglasses will also do.) Place one lens behind the other and look through both. As you rotate one lens and keep the other stationary, you will notice that the intensity of light that comes through varies with the relative orientation of the two lenses. At one point you can see everything, but when the "picket fences" are perpendicular to each other there is total blackness. Actually, by following these instructions you have constructed a simple instrument known as a "polarimeter," central to Pasteur's work and to our story.

As early as 1669, Erasmus Bartholinus, a Danish professor of mathematics and medicine (a combo you don't often find these days), had noted that light passed through a particular form of calcite—a specific crystalline version of calcium carbonate— showed unusual behavior. This behavior was later explained by the notion that the crystal allowed only light waves vibrating in one plane to pass through; in other words, it was a source of "plane-polarized light." Two such crystals could be lined up in such a way that either light passed through both or no light passed through at all. An obvious extension of this experiment was to place substances between the two calcite lenses to see if they affected the path of the plane-polarized light. Some did and some did not.

When the calcite crystals in a "polarimeter" were adjusted so the maximum amount of light was transmitted and a solution of tartaric acid was then placed between them, the light was cut off. But rotating one of the lenses to the right eventually allowed the light to beam through once again. The tartaric acid had rotated the plane of the plane-polarized light to the right! Such substances were termed "dextrorotatory," for "rotating to the right," while those that caused a rotation to the left were labeled "levorotatory." Tartaric acid was clearly dextrorotatory, but racemic acid, which had the same chemical composition, had no effect on the rotation of plane-polarized light. The same effect was noted when these two acids were neutralized and converted into salts. Sodium ammonium tartrate was dextrorotatory and sodium ammonium racemate had no effect on polarized light. This effect is what puzzled Pasteur. What could the difference be?

He began his investigation by examining the crystals of sodium ammonium tartrate and sodium ammonium racemate under a magnifying glass. There was a remarkable difference. While the crystals in the sample of the tartrate were essentially identical, those in the racemate were not. A closer look revealed that the racemate was composed of two kinds of crystals that had a fascinating relationship to each other: they were mirror images. Pasteur laboriously separated the crystals into two piles, dissolved each in water, and placed the samples in a polarimeter. He was amazed to see that one sample rotated the light to the right and the other rotated it by an equal amount to the left. Furthermore, the sample that rotated the light to the right was in every way identical to sodium ammonium tartrate. The puzzle was solved! Since the observation had been made in solutions of the crystals, Pasteur suggested that the molecules in the two types of crystals must also be mirror images of each other. Racemic acid was accordingly seen as nothing other than a mixture of the two mirror-image forms of tartaric acid.

Now back to our Prozac story. Like racemic acid, Prozac (fluoxetine) is also composed of two types of molecules that are mirror images of each other: (R)-fluoxetine and (S)-fluoxetine. They do not behave exactly the same way because the molecules in our body with which they interact also have a certain "handedness." Indeed, (R)-fluoxetine is metabolized more quickly and the (S)-fluoxetine version is thought to be responsible for more side effects. Eli Lilly's plan was to come up with a synthetic procedure that would produce only (R)-fluoxetine, which would then be classified as a new drug and could receive patent protection. The company worked the synthesis out, but since (R)-fluoxetine was metabolized more quickly, it had to be tested at a higher dose than the one found in Prozac. Unfortunately, this higher dose produced side effects on the heart and the plan had to be abandoned. The failure was a major scientific and economic setback.

The concept of marketing a single version of a drug that occurs in mirror-image forms, however, is a good one. Usually one of the "enantiomers," as these mirror-image compounds are called, is more active than the other, while both may be responsible for side effects. Therefore, single-enantiomer preparations can be more effective and produce fewer side effects. This is indeed the case for one of the most widely prescribed drugs in the world, the antiulcer medication omeprazole (Losec), now marketed as an enantiomeric mixture. Here the (S) version is the active one, and Losec's manufacturer, AstraZeneca, has introduced a single-enantiomer version esomeprazole (Nexium). Clinical studies have shown this version to be more effective, with reduced side effects. And its success is due to the ingenuity inherent in separating molecules that are mirror images of each other, ingenuity first demonstrated by Louis Pasteur.

Pasteur's contribution to chemistry is remembered in the elaborate mausoleum at the Institute Pasteur in Paris, where the great scientist is buried. Mosaic tiles on the tomb commemorate various aspects of Pasteur's research. Visitors are likely to recognize that a flock of sheep represents his work on the anthrax vaccine and that a dog symbolizes his conquest of rabies. Undoubtedly, though, many are puzzled by the phrase "une dissymetrie dans les molecules," which is featured on the ornate tableau. But now you know the secret that lies behind these words.

THE LUNATICS

William Withering was a lunatic. But he wasn't crazy. James Watt, perhaps the greatest of British engineers; Joseph Priestley, the discoverer of oxygen; and Josiah Wedgwood, the creator of the famed pottery, were lunatics as well. But they were all perfectly sane. In fact, they were more than sane. These great men were members of the world's first "think tank," the Birmingham Lunar Society. Starting in 1766, this brilliant collection of scientists met once a month to ponder the scientific issues of the day. The meetings were always at the time of the full moon so members could readily find their way home in the darkness. Eventually the date became less important because one of the founders of the Society was William Murdoch, the inventor of coal-gas lighting! The "lunatics" made numerous contributions to society, with Withering making his mark in drug development. In fact, the sophisticated research path used by the modern pharmaceutical industry can be traced back to the dogged attempt by this British country physician to find an answer to the problem of dropsy, perhaps the most common cause of death in the 1700s.

Dropsy occurs when the heart becomes too weak to pump blood around the body. Today the condition is referred to as "congestive heart failure" (CHF), but at the time it was not recognized as a heart problem. The only thing physicians knew was that patients' bodies puffed up; a tremendous paunch characterized the disease. Sometimes arms and legs swelled so much that they were rendered immobile, and fluid filled the chest cavity, impairing lung function. Since the condition prevents the heart from pumping blood around the body effectively, the blood backs up and fluid from it leaks into the surrounding tissues, which is what causes the swelling. The condition usually shows up first as unusual fatigue and shortness of breath, followed by rapid weight gain due to fluid retention. There are many possible causes for CHF. Clogged coronary arteries can weaken the heart, as can years of uncontrolled high blood pressure. Abnormal heart rhythm, valve problems, birth defects, and even alcohol consumption can be predisposing factors. In Withering's day there was nothing that could be done for these unfortunate patients. At least not until our hero fell in love!

Dr. Withering had graduated with an MD in 1766 and set up a practice in Birmingham. There, one of his patients was a beautiful young lady named Helena Cooke. Withering was smitten! There were no ethical issues at the time about doctor-patient relationships, so the doctor felt free to pursue his heart. But how? Withering had learned that Helena loved to paint and flowers were her favorite subject. This knowledge gave him an opening. He began to call on the young lady, always bearing different flowers that she could paint. To make an even greater impression, he began to learn about plants so that he could entertain Helena with stories about his gifts. The doctor's efforts were rewarded in more ways than one. Helena agreed to be his wife, and Withering became an expert botanist! In 1775 he was already working on his first major treatise, *A Botanical Arrangement of All Vegetables Naturally Growing in Great Britain*, when he had occasion to make a house call that would go down in history.

Although Withering's practice was in Birmingham, he routinely made the 60-mile trek to the Stafford Infirmary, where he treated poor patients for free. On one of these trips he agreed to drop in on a patient who lived on the way. He quickly recognized that the woman had dropsy and told her there was, unfortunately, nothing he could do. Much to his surprise, when he looked in on the patient again, he found her decidedly improved. What had she done? It was at this point that Dr. Withering learned about the "wise woman of Shropshire" who had a secret formula for treating dropsy. He sought her out and discovered she was a font of folkloric medications. The dropsy remedy, he was told, was a mixture of some twenty different herbs. Since Withering already knew a great deal about plants, he immediately recognized that the most likely active ingredient was an extract of foxglove. After all, it had been mentioned here and there in herbals over the years as a remedy for ailing hearts.

Withering immediately became interested in trying foxglove on his patients, but in a historical first, he began a systematic investigation to develop a standardized, reproducible form of the plant. First, he determined that the other ingredients in the Shropshire recipe were quite useless and even found that effectiveness depended on when the leaves were harvested and at what temperature they were dried. Then, using 163 patients in what amounts to the first controlled drug trial in history, he found that dried, powdered leaf of foxglove given orally in the proper dose was effective in treating dropsy. However, even a slight overdose could result in catastrophic headaches and vomiting.

"Digitalis," as the preparation came to be called after the fingerlike shapes inside the flower, eventually became the standard treatment for the symptoms of dropsy. By the early twentieth century, the active ingredients digitoxin and digoxin had been isolated, making standardized preparations readily available. Digitalis didn't make patients live longer, but it surely increased their quality of life. Today, with the proper use of angiotensin-converting enzyme inhibitors such as enalapril or captopril and beta blockers such as carvedilol, congestive heart disease patients can actually increase their life expectancy. Digitalis is still used, particularly with patients who suffer atrial fibrillation, a form of irregular heartbeat. Paradoxically, recent research has shown that digitalis therapy is more effective in men than women. There have also been some interesting developments in a nutritional approach to congestive heart disease. Dietary supplements such as taurine, L-carnitine, and coenzyme Q10 may help alleviate the symptoms.

Over 200 years ago, William Withering recognized the difficulty of working with natural products. He was acutely aware of potential toxicity and realized that the effectiveness of his preparations depended on the variety of the plant, storage conditions, temperature at which the extractions were carried out, and, of course, dosage. This simple country doctor would surely be shocked by today's popular slogan "Botanicals are safer." Let me play the role of a modern lunatic and suggest that Withering's 1785 *An Account of the Foxglove* should be required reading for all who believe that natural remedies are always superior to synthetic drugs.

Heady Times at the Bakery

They only used about 3 kilograms of poppy seeds per week, so why, the baker wondered, was his partner ordering a 25-kilogram bag every week? The matter remained a mystery until the partner suffered a seizure, followed by delirium and hallucinations. Hospital tests revealed a staggeringly high level of morphine in the man's blood. Apparently, the baker had been addicted to heroin but had successfully completed a recovery program. Ready access to poppy seeds, however, proved too enticing, and he began to brew about 2 liters of tea every day using about 4 kilograms of poppy seeds. The morphine concentration in the tea was high enough to provoke withdrawal symptoms if he abstained for more than a day. His system finally capitulated and he experienced a seizure typical of opium overdose. This event scared him into enrolling in a rehabilitation program, during which his tea-drinking habit was replaced with slow-release morphine (MS Contin), from which he was eventually weaned. Presumably he now specializes in baking muffins.

Our baker was following in footsteps that are at least 6,000 years old. The ancient Sumerians referred to the poppy as the "joy plant" in tribute to its euphoria-inducing properties. Egyptian physicians may have been the first to suggest a medical use for the plant in the famed *Ebers Papyrus,* written between 3000 and 1500 B.C. This document is a compilation of medical writings featuring everything from diseases of the tongue to diseases of the toes and offers a remedy for a colicky child that can be made from "pods of the poppy plant and fly dirt which is on the wall." One would assume that the active ingredient was the opium from the poppy and not the fly excrement. The ancient Greeks also made their contribution to poppy scholarship. Dioscorides, in the first century A.D., wrote a scientific thesis on opium, describing its collection from the unripe pods of *Papaver somniferum,* the opium poppy. Specially sharpened knives were used to score the pods, which then oozed a milky white substance that was dried to a powder. This is essentially the technique still used today on the poppy plantations of Afghanistan, India, China, and Pakistan.

Why is opium produced on such a massive scale? For two reasons. First, morphine and codeine, two of the prime painkillers used in modern medicine, are extracted from opium. Second, a huge black market unfortunately exists for heroin, which is synthesized from morphine. Morphine is the proverbial double-edged sword. It probably has both alleviated and caused more misery than any other chemical in history. The beneficial side of morphine was quickly noted in Europe after Paracelsus, the noted Swiss alchemist, introduced laudanum, an alcohol extract of opium. It became a staple treatment for diarrhea, pain, and insomnia. Queen Victoria used it, and numerous children in the Victorian era were put to bed with Godfrey's Cordial, which contained a hefty dose of opium. Everyone who had experience with opium understood why Friedrich Wilhelm Serturner, who had isolated the active principle in 1906, named it "morphine" after the Greek god of dreams: it not only alleviated pain and induced sleep but also was capable of triggering an euphoric, dream-like state. Thomas de Quincey summed the situation up effectively in 1821 in his *Confessions of an English Opium Eater.* "Here," he said, "was a panacea for all human woes, here was the secret of happiness . . . portable ecstasies might be corked up in a pint bottle." As it turned out, that bottle held misery as well as ecstasy.

"Morphinomania" seized England as fashionable ladies held "morphine parties," during which they injected each other with morphine using quaint, specially designed syringes. By this time it had become apparent that smoking opium was addictive, but the prevailing theory was that injected morphine did not have this effect. Wrong! The invention of the hypodermic syringe in 1853 was, in fact, a huge contributor to the addiction epidemic. Soldiers who had been treated for pain with morphine—in many cases in a reckless fashion—returned from the Franco-Prussian and American Civil Wars as addicts. Amazingly, one of the first attempts to counter morphine addiction relied on using heroin as a substitute—the same heroin that was destined to become the most notorious drug of our time.

Heroin does not occur in nature. It was one of the first morphine derivatives synthesized when researchers began a process—which continues to this day—of trying to modify the molecule with a view toward improving its painkilling effect and reducing its addictive potential. In 1874, Charles Alder Wright at St. Mary's Hospital in London, England, reacted morphine with acetic anydride and produced diacetylmorphine, a compound that eventually came to the attention of the Friedrich Bayer Company in Germany. In 1898, Bayer's Felix Hoffman had solved the problem of stomach irritation associated with the analgesic salicylic acid by treating it with acetic anhydride to produce aspirin. He thought that such an "acetylation" might also improve the properties of morphine and reduce its common side effects of nausea, vomiting, and respiratory depression. Early studies by Bayer showed that the novel compound had some tantalizing possibilities. It seemed to clear the lungs of phlegm and suppress coughs in a "heroic" fashion. Accordingly, Bayer christened diacetylmorphine "heroin" and began to market it as a cough suppressant, often advertising it on the same bill as aspirin.

By about 1911 it had become apparent that Heinrich Dreser, head of Bayer's pharmacology lab, had erred about heroin's effects on respiration. Like morphine, it actually depressed breathing. If not for this mistake, Bayer probably would never have marketed heroin, as it offered no other advantage over morphine. It is possible, then, that we would have avoided being saddled with the huge social problems associated with this drug.

Morphine and its various synthetic derivatives, such as oxycodone (OxyContin), are still the mainstays of pain treatment. Chemists have not yet been successful in producing analogs that retain morphine's painkilling effect while eliminating its side effects and potential for abuse. Unfortunately, this quandary has sometimes resulted in the underutilization of this highly effective painkiller. It is certainly possible to use morphine effectively and still maintain vigilance against its abuse. However, sometimes even this vigilance goes overboard. For example, it is understandable that paroled prisoners have to undergo tests for drug use. But what about the case of the Florida parolee who swore he had used no drugs yet was almost sent back to jail because he tested positive for morphine? As chance would have it, he had eaten a poppy seed bagel for breakfast and the sophisticated test now in use picked up traces of morphine in his urine! The former convict has now developed a fondness for sesame seed bagels instead.

Newton and Mercury

You know about his falling apple. You know about his prism and its rainbow. You may even know that some surveys have declared him the second most influential person of all time, ranking in between Mohammed and Jesus. But I suspect few of you know that Isaac Newton had a passion for chemistry and spent about thirty years of his life among the flasks and beakers of his laboratory near Cambridge in the pursuit of, well, nobody really knows. Newton kept extensive notes but never formally published anything about his chemical investigations, probably because at the time such activities were frowned upon. The royals feared that if an alchemist discovered an easy way to make gold, the country's monetary system would be destroyed. On this account, they had nothing to fear from Newton. His genius does not appear to have extended to chemistry.

Simply stated, genius is seeing what everyone else sees but thinking what no one else thinks. By this credo, or indeed by any other, Isaac Newton was a genius. He saw that apple fall and surmised that the force that attracted it to the earth was the same one that held the moon in orbit around the earth. He then went on to formulate the laws of motion and indirectly inspired the science of space travel. Newton's Third Law—for every action there is an equal and opposite reaction—is confirmed each time a rocket is launched into space. His use of a prism to separate white light into the colors of the rainbow laid the foundations for the science of optics. His *Mathematical Principles of Natural Philosophy*, published in 1687, is undoubtedly one of the single most important works in the history of science. But there was another side to this great man.

Isaac Newton, by all accounts, was not a pleasant fellow. As a youngster Newton had battled with his stepfather and even threatened to burn his house down. He hated the man with a passion for forcing him to live with his grandmother, which meant the young Newton had been separated from his mother. Later in life, Newton showed definite psychotic tendencies characterized by occasional withdrawal from human contact, fits of anger, and periods of depression. His battles with other scientists over the details of his theories were ferocious. Let's just say that Newton did not take criticism well.

The last twenty-five years of his life were dominated by a vicious feud with German mathematician Gottfried Leibnitz over who had been the first to formulate the concepts of calculus. Historically, it seems that Leibnitz had independently arrived at his ideas several years after Newton but was the first to publish. The two squabbled and brawled in the pages of scientific publications, with Newton carrying on the battle even after Leibnitz had died. On hearing of Leibnitz's passing, the great physicist reportedly expressed his pleasure at "having broken the German's heart."

Most historians have ascribed Newton's bouts of bizarre behavior to deeply rooted psychological problems, perhaps brought about by the trauma of being separated from his mother during his formative years. But there is another theory, and it brings us back to Newton's chemical exploits. According to his notes, Newton began to dabble in alchemy around 1687 after reading the works of George Starkey, an American who wrote under the pseudonym *Eirenaeus Philalethes,* which translates into "peaceful lover of truth." Starkey was educated at Harvard, where he was introduced both to alchemy and medicine and spent his life looking for the "universal remedy." This was some vaguely defined potion that could change substances from one form into another and cure disease.

Starkey believed that the ancient Greeks and Romans had discovered the secret and encoded the procedure in mythology. He was particularly taken by the story of Vulcan, the husband of Venus, who caught his wife in a delicate situation with Mars. Vulcan didn't like being a cuckold one bit, so he captured the cavorting lovers in a fine metal net and strung them up for all to see. In alchemy, Venus stood for copper, Mars for iron, and Vulcan for fire. Starkey "interpreted" the mythological story and heated a mixture of copper, iron, and stibnite (antimony sulfide) to create a beautiful purple copper-antimony alloy he called "The Net." It wasn't clear what The Net was supposed to do, but it certainly interested Newton, who recorded exact instructions for its production.

Chances are that Newton believed The Net to be a key substance in alchemy, perhaps needed to turn metals such as mercury into gold. At this point the account gets really interesting because Newton's notebooks describe how he heated mercury compounds and tasted concoctions brewed with the element. And it is during this period of Newton's life that he began to show mental disturbances and started to suffer from sleeplessness and auditory hallucinations, both of which are symptoms of mercury poisoning! Unfortunately, if Newton was indeed poisoned by mercury, it was for naught. As far as we can tell, his chemical exploits never amounted to anything significant.

So do we look to Oedipus or mercury for an explanation of Newton's behavior? A tantalizing clue emerges from a lock of hair kept at Trinity College in England, a lock that supposedly belonged to this eminent man. Analysis of the hair has shown a mercury level of about 200 parts per million, compared with a normal level of roughly 5 parts per million. But is it really Newton's hair? We will never know because exhumation of the first scientist to be buried in Westminster Abbey is, of course, out of the question. While there may be doubt about the mercury poisoning, there is no doubt about Newton's pivotal role in the development of science. This certainty was perhaps best expressed by the poet Alexander Pope, a contemporary of Newton's: "Nature and Nature's Laws lay hid in the night; / God said, Let Newton be! and all was light!" And just imagine what Newton might have accomplished if he had left chemistry alone and stuck to physics!

FROG LEGS AND A GIANT LEAP

Sometime around 600 B.C., and for reasons known only to himself, Thales of Miletus, a Greek mathematician and philosopher, rubbed a piece of amber with animal fur and made an astounding discovery. The amber (basically fossilized tree sap) became a virtual magnet for feathers and small particles of dust. Thales didn't understand what was going on, but we do. Electrons were transferred from the fur to the amber, giving it a negative charge. This charge then repelled electrons from the surface of the feather toward its interior, giving that surface a positive charge. The attraction between the positive and negative surfaces then resulted in the observed effect. An interesting footnote to this story is that the ancient Greek word for amber is *elektron*, a

term scientists later appropriated to describe the smallest particle of negative charge.

Thales's observation remained a scientific curiosity until the seventeenth century, when it sparked the interest of Otto von Guericke, a German scientist. He constructed a large sphere made of sulfur and mounted it on an axle so that it could be rapidly rotated by means of a handle. Placing a hand on the sphere resulted in the transfer of electrons, giving the sphere a negative charge. Von Guericke had constructed the world's first static electricity generator! He could think of no real use for the contraption and it was pretty well relegated to being a parlor game: a gentleman standing on a wooden stool could be charged, and when he kissed a lady standing on the ground, the sparks would literally fly.

Luigi Galvani, an Italian professor of anatomy, became intrigued by von Guericke's machine and decided to investigate the effects of static electricity on biological systems. One day in

1780 he was looking for observable changes in a dissected frog as an assistant cranked an electrostatic generator nearby. Nothing much happened until Galvani picked up a steel scalpel. As he was cutting away, a spark suddenly jumped from the machine to the scalpel, and to Galvani's amazement, the frog's leg began to twitch. It was as if the animal had come back to life!

If a small spark could cause such an effect, what could a large one do? Well, Galvani knew that lightning was just one big spark. So, during a thunderstorm, he strung some freshly dissected frog legs on a brass wire that he attached to the iron fence in front of his house. Then he waited for lightning to strike. It didn't. But something else did happen.

The frog legs swayed back and forth in the breeze, and every time they touched the iron fence they twitched uncontrollably. Now this was really strange! Muscle stimulation in response to electricity was one thing, but these apparently spontaneous spasms were something else altogether. Galvani had to come up with a rationale for the bizarre event. He hypothesized that somehow, electricity was stored in the frogs' muscles and it could be released under certain conditions. "Animal electricity," he called it.

The idea of animal electricity did not sit well with Alessandro Volta, Galvani's countryman and a professor of physics at the University of Pavia. He searched for an alternate explanation for the frog legs' activity in the thunderstorm and soon found it. The electricity had not come from the frogs; it had come from the brass support and the iron fence! Volta discovered that an electric current could be generated when two dissimilar metals, such as iron and brass, were connected through a moist conductor, which in this case was frog parts. As the frog legs attached to the brass wire dangled in the wind, a current flowed whenever they touched the iron fence. The Italian physicist had invented the battery!

Volta soon found that silver and zinc formed an even better set of metals for generating electricity and that frog legs could be replaced by a piece of moist cardboard. He went on to build little sandwiches of silver and zinc disks separated by wet cardboard and stacked them one on top of the other to create a "Voltaic pile." He noted that grasping the pile between the hands had the same effect as holding an electric eel.

News of Volta's discovery electrified Europe. The Italian scientist was invited to speak at London's Royal Society and was even asked to give a private demonstration of the dancing frog legs to Emperor Napoleon. Soon scientists across Europe were exploring the potential uses of Volta's discovery, and by 1807, Humphry Davy in England had built a 2,000-cell battery, which he connected to a wire that became hot and glowed. This crude prototype of an incandescent light was therefore a direct descendant of Galvani's dancing frog legs!

Why those frog legs danced remained a matter of mystery until the early 1900s, when physiologists concluded that the nervous system was akin to a vast, complex grid of wires through which tiny electrical currents flowed. Nerve impulses could therefore be thought of as electrical signals that traveled along these "wires" and stimulated muscles to move, lungs to breathe, and the heart to beat. Just how the electrical signal was transformed into the required physical action was not known.

Then, in 1920, Otto Loewi, a German physiologist, performed a seminal experiment. Once again frogs took center stage. Loewi used pulses of electricity to stimulate a nerve that he knew would cause a frog's heartbeat to slow down. With a syringe, he then removed a small amount of fluid from the area where the nerve met the heart and applied this fluid to the heart of another frog. To his amazement, the second heart also began to beat more slowly! This observation meant that electrical stimulation must have released a chemical capable of triggering physiological activity. That chemical eventually turned out to be acetylcholine, the first neurotransmitter ever to be isolated. The discovery of many other neurotransmitters, some with household names such as adrenaline and serotonin, soon followed. Understanding the activity of these compounds has led to the development of medications designed to treat ailments ranging from heart disease to depression. And it all was sparked by a spark—and frog legs.

ANIMAL, VEGETABLE, OR MINERAL?

The St. Louis World's Fair of 1904 featured a most unusual exhibit. A bright spotlight illuminated a small vial that contained some nondescript white crystals. But these were no ordinary crystals. They were crystals of synthetic urea, crystals that had changed the course of chemical history.

Organic chemistry is a most wondrous endeavor. Imagine creating a new painkiller, a novel anticancer drug, or a revolutionary plastic by manipulating substances too small to be seen. These invisible building blocks of matter, these "molecules," govern every aspect of our lives, yet few people have a grasp of what they really are. Of course, one does not have to understand molecules in order to manipulate them. Throughout history people have made soap from animal fat, smelted metals from ores, brewed alcohol from grapes, and extracted dyestuffs from plants without knowing anything about molecules. It really wasn't until the nineteenth century that a comprehensive picture of the fundamental substances that constitute matter began to emerge. Antoine Lavoisier had laid the foundation with his clear description of how elements combined to form compounds, and John Dalton had introduced the idea that the elements themselves were composed of little particles he called "atoms."

But there was a complication in the quest to explain the properties of matter in terms of atoms. Substances derived from living organisms were somehow very different from those found in nonliving matter. They were more complex, more difficult to separate, and were usually destroyed by heat. A green ore could ooze metallic copper when heated, but under the same conditions a flower would be quickly converted to a piece of useless charred matter. Clearly the behavior of these natural materials required specialized study. "Organic chemistry" was the term suggested in 1808 by the Swedish chemist Joens Jacob Berzelius for the new discipline that would be dedicated to the study of substances from living organisms.

It turned out to be a difficult discipline. Although most organic materials were found to contain only carbon, oxygen, hydrogen, and nitrogen, they defied any attempt at synthesis from these elements. Some scientists, including Berzelius, suggested that a "vital force" was inherent to organic substances, a force that could only be produced by living things. According to this theory, there was no point in trying to make quinine in the laboratory because the vital force that gave this substance its properties could only be infused into it by the living cinchona tree. Organic chemistry seemed very complex, and Friedrich Wohler, a leading German chemist in the early 1800s, summed up his frustration succinctly: "Organic chemistry nowadays almost drives me mad. To me it appears like a primeval tropical forest full of the most remarkable things, a dreadful endless jungle into which one does not dare enter, for there seems no way out."

But there was a way out. And it was Wohler who found it. He had originally trained as a physician and had isolated urea during an investigation of the waste products found in urine. This study captured his imagination and Wohler decided to focus on chemistry instead of medicine. He traveled to Sweden to study under Berzelius, who was already established as a chemical luminary thanks in part to his creation of the letter symbols used to represent the elements, the symbols we still use today. It was in Sweden that Wohler heard about the impassable gulf separating organic substances from inorganics.

Wohler went on to become professor of chemistry at Gottingen, where he became interested in substances that could release cyanide when heated. One day in 1828, he heated up some ammonium cyanate expecting to liberate some cyanide. None was released. But his original crystals had taken on a new form. Their weight had not changed, but these crystals had a different melting point and a different appearance from his starting material. What could have happened? When Wohler carefully examined the crystals, he noticed that their shape seemed familiar. Where had he seen them before? His mind darted back to his medical school days and then he knew—he was looking at crystals of urea!

Wohler excitedly wrote to Berzelius: "I must tell you that I can prepare urea without requiring a kidney of an animal, either man or dog." He somewhat remorsefully added that he had witnessed "the great tragedy of science, the slaying of a beautiful hypothesis by an ugly fact." Indeed he had. On that day in 1828, the "vital force" theory was mortally wounded, although it would linger for a few more years. Friedrich Wohler had made an organic substance in the laboratory from an inorganic one. Clearly, in their attempts to make novel substances, scientists would be limited not by any vital force but by their ingenuity.

Although Wohler had shown that there was nothing mystical about organic materials, neither he nor his contemporaries could explain why they behaved differently from inorganic substances. It had something to do with their chemical makeup, but what exactly? There was a clue lurking in Wohler's previous exploits.

While studying with Berzelius, he had carried out a chemical analysis of silver cyanate and found it to be composed of silver, oxygen, nitrogen, and carbon. But another young German chemist, Justus von Liebig, had just published an analysis of silver fulminate, a completely different substance, which also was made up of the same elements in exactly the same weight ratio as Wohler's compound. Liebig called Wohler incompetent and claimed that his analysis must be wrong. But when the two finally met and went over their data, they agreed that they were both correct. How could this be?

Berzelius stepped in and suggested that both substances could be made of the same elements yet still be different if their atoms were joined together in different ways. Eureka! Now Wohler's urea experiment also made sense. Why did the original ammonium cyanate and the final urea have the same weight? Because the only difference was the arrangement of their atoms! It was now becoming clear that the properties of materials reflected not only what kinds of atoms they were made of but also how these atoms were arranged. The richness and complexity of organic chemistry lay in the fact that atoms of hydrogen, oxygen, carbon, and nitrogen could combine in numerous ways to generate a myriad of molecules.

Liebig and Wohler couldn't even dream of how these atoms united to form molecules. Luckily, August Wilhelm Kekule could. Kekule had planned to study architecture until he heard one of Liebig's lectures on chemistry. He was hooked. Then and there he decided to devote himself to solving the mysteries of atoms and molecules. He finally managed to do just that, but not at a lab bench. The revelation supposedly came to him in a dream. Kekule had been riding a London omnibus when he dozed off. He later described how "atoms began to gambol in front of his eyes and suddenly one latched on to another, then another and quickly a chain of atoms was formed!" At this moment he was roused by the cry of "Clapham Road," but the basics of molecular structure had been forged in his mind.

Kekule concluded that organic compounds were very complex and had great diversity because carbon atoms were able form bonds to each other, creating an almost infinite array of patterns. In order to rationalize the ratios of the weights of elements in organic compounds, Kekule had even postulated that each carbon atom could form four such bonds. A Scot, Archibald Couper, put the finishing touches to this theory of molecular structure by drawing the first molecular diagrams, in which the atoms were represented by their letter symbols and bonds by straight lines. Organic chemistry was seen as not only the study of substances from living sources but also the study of the compounds of carbon. The field was thrown open. Organic chemists now knew what molecules were and soon found ways to make them. And make them they did. By the millions. Drugs, dyes, plastics, and cosmetics soon burst from their test tubes.

Although molecules as complex as DNA can be synthesized today, organic chemists still have a special appreciation for urea. Indeed, upon seeing Wohler's original urea sample at the St. Louis World's Fair, an American chemist remarked that he felt like a pilgrim "who has just seen a piece of the True Cross."

HORROR FILM

There are some images that become indelibly etched on one's mind. Like a giant spider crawling across an immense web. Or a boy flying through the air, hanging on to a genie's fluttering pigtail. These pictures dance in my mind today almost as vividly as when I first saw them projected on a screen in Hungary way back in 1954. *The Thief of Baghdad,* in glorious color, was the first movie I ever saw. What a feast for the eyes! The flying carpets, the swordfights, the teeming markets—they whetted my appetite for more. So you can appreciate my disappointment when we moved to Canada in 1956 and I was told that I wouldn't be going to the movies.

What a strange country Canada was! Why weren't children allowed to experience all those swashbuckling adventures that could be captured on film? As I later found out, the ruling had nothing to do with what was on the film; rather, it had to do with what was in the film. In 1927, a fire had broken out in the projection booth of the Palace Theatre in Montreal. The flames spread quickly, and seventy-eight children perished in the inferno. The culprit was thought to be cellulose nitrate film, which had ignited from the heat of the projector's bulb. Within a year, a law was passed prohibiting children from going to the movies.

The roots of this tragedy can be traced back to early attempts to simplify photography. At first, the silver compounds that produced an image when exposed to light were spread onto either cumbersome glass plates or delicate paper. These materials were difficult to work with, particularly for amateurs. Then, in 1885, American George Eastman found a way to apply the light-sensitive crystals to a transparent plastic film that could be readily rolled up and used in simple cameras. Eastman's discovery hinged on the earlier contributions of Friedrich Schönbein in Switzerland and John Wesley Hyatt in the US. Schönbein had noted that treating cellulose, which he obtained from cotton, with a mixture of nitric and sulfuric acids yielded highly flammable nitrocellulose. When this substance was dissolved in alcohol, and the alcohol was allowed to evaporate, a plastic film termed "collodion" was left behind. Hyatt, searching for a substitute for the ivory needed to make billiard balls, mixed collodion with camphor and produced the world's first commercially viable plastic, called "celluloid." Celluloid is what Eastman coated with silver bromide to produce his first flexible roll film. It immediately captured the imagination of the public, and perhaps more significantly, it caught the attention of America's greatest inventor.

Thomas Edison immediately recognized the potential offered by the film's flexibility. By 1889 he had parlayed Eastman's discovery into his Kinetoscope, a sort of "peep show" device that allowed people to look through a lens to view a moving filmstrip. Although Edison considered the possibility of projecting the moving pictures, he thought more money could be made with individual Kinetoscopes lined up in special Kinetoscope parlors, with each one offering a different film. The idea was one of his few mistakes. The mistake would not be repeated by the Lumière brothers in France, who got excited after seeing a Kinetoscope demonstration in Paris. Before long they devised a "cinematograph," and on December 28, 1895, in the basement lounge of the Grand Café on the Boulevard des Capucines in Paris, they treated a gathering to "moving pictures." Of course, by today's standards it was not a great cinematographic extravaganza, but some members of the audience jumped up and ran when a train appeared. They thought it was going to burst through the screen!

The train didn't burst through the screen, but far too often, flickering flames appeared to do just that. Sometimes in the middle of a movie, the screen seemed to catch fire as the cellulose nitrate film was ignited by the heat of the projector bulb. After a number of disastrous fires, Eastman developed "safety film." This film used cellulose acetate, a far less flammable material made by treating cellulose with acetic acid. By 1948 acetate had become the industry standard and fear of movie theaters going up in smoke dissipated. Furthermore, cellulose acetate appeared to solve another problem that had plagued nitrate film. Over time, nitrate film slowly reacted with oxygen and underwent a slow "burn." Within a couple of decades, it yellowed and became brittle. Cellulose acetate didn't burn, but as it turned out, it was susceptible to a different kind of decay. As the film aged, it released acetic acid, which slowly broke down the cellulose, sometimes eating holes right through the film. This decay has been termed the "vinegar syndrome." Anyone who opens an old film canister and experiences the olfactory assault will immediately understand why. In response to this problem, polyester film was introduced in 1960. It doesn't degrade readily but is harder to splice and therefore to edit. Today, both cellulose acetate and polyester films are used; appropriate storage conditions with controlled temperature and humidity ensure that they will last for many years. Edison's classic *Great Train Robbery* has been stored in a refrigerator for decades and is in great shape.

The same cannot be said of original copies of *Casablanca,* which was filmed on celluloid. But don't worry—the film classics will not be lost; virtually all older movies have been transferred onto newer films or videotape. But videotape doesn't last forever either. The metal oxide particles that form the image, as well as the polyurethane binder that holds them in place on the polyester base, degrade with time. This decomposition may not matter that much because movies are now being transferred to DVDs. The molded polycarbonate that is encoded with the pits that reflect the laser reader's light beam is very stable and will probably last a few hundred years.

Today, with the flammability problem licked, children are allowed to go to the movies and we worry more about what's on the film than what's in it. Had I risked my life when I went to see *The Thief of Baghdad*? No. As I've since learned, color movies were always filmed on cellulose acetate, never on celluloid. These memories stimulated me to try to recapture that little slice of my youth by ordering a copy of *The Thief of Baghdad.* On DVD, of course.

ABSINTHE MAKES THE TART GROW FONDER

I was gazing out the bus window as we rambled through Notting Hill, heading toward Heathrow Airport in London. Images of Hugh Grant and Julia Roberts strolling through those streets were flashing through my mind when I was caught off guard by a huge molecule. Or, I should say, a huge *model* of a molecule. It was sitting in a shop window right under a gigantic neon sign that read "Pharmacy." Now that's just the kind of thing to grab any chemist's attention. But all I could do was grab my camera and take a picture. I wished I had seen the sign earlier and visited the place—after all, how often does one get the chance to mingle with a giant molecule?

Developing the film produced a surprise. "Pharmacy" was not a pharmacy at all. It was a bar! And a rather avant-garde one, as it turned out. A quick check on the Web revealed that in 1998 it became the first bar in England to serve absinthe since the drink had disappeared from the country's shelves in the early 1900s. Noting its return stimulated me to indulge in the lore and science of absinthe.

The "Green Fairy," as the beverage was known in its heyday, was first produced by Henri-Louis Pernod in 1797 after he purchased the recipe from his father-in-law, Major Henri Dubied. The green beverage was brewed from a mixture of herbs, including anise, fennel, nutmeg, and juniper. But the key ingredient was an extract of *Artemesia absinthium,* a shrub better known as wormwood. This plant had received a fair degree of interest during the Middle Ages, when its oily extract was used to treat intestinal worms, although not with any documented efficacy. Nobody really knows how or why Major Dubied thought of putting the bitter extract into a beverage, but the best bet is that he used it as a medicine for his soldiers. Something that tasted so bad had to be good for something.

Apparently it was. Soon soldiers were reporting unusual mental clarity and an increased libido. That was all the encouragement artists needed, and they began to investigate the effects of absinthe. During "l'heure verte," or the green hour, the likes of Paul Gauguin, Henri de Toulouse-Lautrec, and Vincent van Gogh practiced the ritual of placing a sugar cube on top of a special perforated silver spoon and dripping water over it into a glass of absinthe until the green beverage turned a cloudy white. There's some interesting chemistry going on here. The plant materials used to flavor absinthe contain compounds called "terpenes," which are soluble in alcohol but not in water. As water is slowly added, they precipitate out and result in the characteristic milkiness associated with absinthe. Today, Toulouse-Lautrec's absinthe spoon is prominently displayed in the Absinthe Museum in Auvers-sur-Oise, France, the little village where van Gogh is buried.

Artists claimed that absinthe unleashed their creativity. Unfortunately, it unleashed other things as well—such as hallucinations, convulsions, and bizarre fits. Thujone, a specific terpene found in wormwood, was blamed for these reactions, although it is not likely that it was the culprit. While thujone can cause nerve cells to fire at an unusual rate by blocking receptors for the inhibitory neurotransmitter gamma-aminobutanoic acid (GABA), it is doubtful that there was ever enough of the substance in absinthe to cause the reported psychoactive effects. The truth is that absinthe was at least 75 percent alcohol by volume (150 proof) and it was cheap. It was easy to overindulge. The nasty effects were probably due to the alcohol. In some cases, cheap absinthe was colored with copper compounds and fortified with methanol, both of which can produce toxic reactions. Thujone, however, cannot be absolved of all blame. In high doses it can certainly cause problems. A well-documented recent case describes kidney failure in a man who consumed some "oil of wormwood" he had purchased from an "aromatherapy" provider over the Internet. Thinking he had purchased absinthe, he downed the contents of the little bottle. Luckily, he survived.

Absinthe's popularity in France increased as wine became more and more expensive thanks to repeated failures of the grape crop, the result of infestation by tiny insects known as "phylloxera." Mass inebriation also increased and cries to ban the drink began to be heard. Then came a pivotal moment. In 1905, Jean Lanfray, a Swiss farmer, murdered his entire family after an "absinthe binge." Newspapers across Europe picked up the story, failing to mention that the man had also consumed a few bottles of wine and several shots of brandy. This was just the kind of account the abolitionists needed to apply political pressure, and by 1914 absinthe had been made illegal almost everywhere in the Western world. Almost everywhere.

Spain and the Czech Republic kept producing the Green Fairy, although with very little wormwood extract. Then, in 1998, a couple of clever Britons realized that absinthe had never actually been made illegal in England, and they decided to import it. The drink took off. Absinthe's history and mystique were enough to make young bargoers shell out about nine dollars for a shot. Most just experienced the effects of a high dose of alcohol, but one young man said that absinthe did make him do strange things. Nothing like cutting off his ear à la Van Gogh; he just developed an urge to take a chicken from the refrigerator and beat it with a hammer.

In Canada, the provinces of Ontario, British Columbia, and Quebec have decided to allow the sale of absinthe. This could turn out to be profitable move because people like the idea of dipping into something that is illegal elsewhere. However, I don't think Canadians have to worry about any thujone effects. The amount allowed in the beverage is minimal and is strictly regulated. I myself have sampled some Hill's Absinthe brought back from the Czech Republic. To be frank, it tasted like bitter mouthwash with an alcoholic kick. But I felt absolutely no urge to pound a chicken. Too bad. I like chicken schnitzel.

Pain Isn't a Laughing Matter

London was not a fun city in 1824. Smoke filled the air, open sewers ran through the streets, and squalor and illness were everywhere. So it wasn't surprising that the public responded enthusiastically to a playbill advertising a performance that would "Let those Laugh now, who have never laughed before, / And those who always laugh, now Laugh the more." Ladies and gentlemen lucky enough to secure a ticket were treated to quite a show. Mr. Henry, as the conjurer called himself, strode out onto the stage and began by turning wine into water and the water back into wine. A cup of coffee magically changed into rice, and a dove placed in a box disappeared only to reappear across the stage. The magician's feats became more and more impressive as the show built toward its climax. Finally, the curtain rose to reveal a stage devoid of the usual magic paraphernalia. Only a solitary table bearing a few inflated pig bladders basked in the spotlight.

Mr. Henry explained to the audience that the bladders were filled with a gas that had been discovered some fifty years earlier by the famed English chemist Joseph Priestley, who had made it by heating ammonium nitrate. "Nitrous air diminished," was Priestley's name for the gas, which eventually would be called "nitrous oxide" by chemists or just plain "laughing gas" by others. The audience was about to find out why.

The fascinating properties of laughing gas were first recognized by young Humphry Davy, who had been hired to study the therapeutic properties of various vapors at Dr. Beddoes's Pneumatic Medical Institution in Bristol. Beddoes was a firm believer in the treatment of disease by the inhalation of chemicals and had established his institute in 1798 with help from James Watt, the inventor of the steam engine, whose son suffered from tuberculosis. One of Beddoes's ideas was to take cows into the sick rooms of his tubercular patients and have them "purify" the air with their sweet breath. It didn't work. Another of his ideas, though, was destined for a date with history. Dr. Beddoes asked Davy, his young assistant, to investigate the physiological properties of nitrous oxide. By 1800 Davy had accumulated enough information to publish a book on the gas, describing the exhilarating sensations and mirth it could produce. He even noted that inhaling nitrous oxide relieved headaches and toothaches and suggested that physicians consider its use in surgery as an anesthetic. Unfortunately, they didn't.

In 1801 Humphry Davy moved to the Royal Institution of Great Britain to take up a position as director of its chemical laboratory. It was during this tenure that he gained fame as a public lecturer. People flocked to his Friday evening discourses on the latest advances in science. One of his most popular presentations involved the generation of nitrous oxide and its inhalation by volunteers from the audience. This experiment captured the imagination of Mr. Henry, who thought it would be a fitting finale to his conjuring performance. And it was! Audience members were tickled with glee as they watched distinguished gentlemen, under the influence of nitrous oxide, cavort on the stage in the most bizarre fashion. Some volunteers became pugnacious while others laughed and blabbered incessantly, looking around for ladies to kiss. It was a gas!

Nitrous oxide inhalation was also featured in the lectures of traveling speakers in America. One of the best known of these speakers was Samuel Colt, who achieved long-lasting fame as the inventor of the first mass-produced revolver. He raised the money needed to patent his invention with his public exhibitions of laughing gas. Gardner Quincey Colton was another popular science lecturer who lucked into a spot in history on December 10, 1844, in Hartford, Connecticut. One of the volunteers for his laughing gas demonstration was Horace Wells, a local dentist and, as it turned out, an astute scientific observer. Wells noted that Samuel Cooley, another volunteer, bruised himself severely by banging into furniture as he pranced about the stage under the spell of nitrous oxide. But he seemed to be oblivious to the pain. The next day, Wells invited Colton to his dental office and inhaled some of the nitrous oxide he had asked the showman to bring along. Then he motioned to Dr. Riggs, his colleague, to pull out an aching tooth that had been bothering him. Wells felt no pain, and on that day a new era of painless dentistry was born. Dentists and physicians still use nitrous oxide to dull the pain of patients who rarely realize the debt of gratitude they owe to a group of select scientists, conjurers, and dentists who plied their trade in the first half of the nineteenth century.

Today, it is not only doctors and dentists who use nitrous oxide, unfortunately. Since the gas affords an escape from reality, it has become an item of abuse, in some cases with tragic consequences. Nitrous oxide is very soluble in fat, and for that reason it is an ideal gas for whipping cream. Next time you're having a banana split, notice the container you use to deliver the whipped cream. It looks like a seltzer bottle fitted with a little cartridge. The cartridge contains nitrous oxide, which whips the cream as it is released. But it can do more than whip the cream; it can kill you. Granted, death is an unlikely scenario and is only possible if a huge amount of the gas is inhaled without oxygen. Still, it has happened.

A twenty-year-old student in Blacksburg, Virginia, was found dead in his apartment surrounded by nitrous oxide cartridges and balloons. Police also found a "cracker," an aluminum tool used to puncture the top of the cartridge so that the gas can be released into a balloon and then inhaled. Apparently, in an attempt to get high on laughing gas, the unfortunate student had asphyxiated himself. The nitrous oxide cartridges had been purchased by mail order from a store in Tempe, Arizona. Agents from the US Food and Drug Administration raided the place and seized boxes of the cartridges, crackers, and punching bag balloons. These items were enough proof that the owner was selling nitrous oxide for purposes of inhalation, even though the boxes were marked for "food use." The owner was fined $40,000 and sentenced to fifteen months in a federal prison. I bet he isn't laughing about it.

BALONEY!

Plain Baloney

If you live in Montreal, Canada, and you've seen a car driving around with a slice of baloney taped to the hood, don't worry—the car was mine. I wasn't trying to set a new fashion trend; I was just trying to find an answer to one of the hundreds of fascinating questions that have come my way. Let me share some of them with you.

The baloney experiment was prompted by a query about the rumor that this cold cut can destroy the paint on cars. I hadn't heard this particular gem before, although I must admit it didn't surprise me. The Web is rife with rumors about virtually everything that exists, be it soybeans, antiperspirants, fabric softeners, canola oil, or aspartame. All these items, we are told, destroy our health. We are also warned about sitting on public toilets (lethal spiders lurk inside the bowls), drinking from pop cans (the rims are tainted by rat urine), using plastic pot scrubbers (they contain poisonous fungicides), and licking ATM envelopes (the glue harbors cockroach eggs, which can hatch on the tongue). Processed meats are said to be veritable poison, loaded with delightful ingredients such as earthworms and eyeballs. Never mind that meat processors, for purely economic reasons, would be crazy to put cow eyeballs into their products; eyeballs fetch far more when sold to research and teaching labs. The paint rumor, I figured, was dreamt up by someone who wanted to emphasize the risks of processed meats: "If it destroys your car's paint, imagine what it does to your insides!"

Now, I'm not going to tell you that such foods offer great nutrition. Baloney is full of fat and salt. There are also some questions about the nitrites used as preservatives. But destroying car paint? That claim seemed far-fetched. How would anyone have made this discovery in the first place? What could have possibly motivated someone to apply baloney to a car? Still, there was nothing in the scientific literature about the subject, so the only way to investigate the allegation was to try it. And I did. I drove around for a day with a slice of baloney taped to my car. When I finally removed the meat at night, I noticed only one effect: the area that had been covered by the baloney was cleaner than the rest of the car. It looked as if it had been waxed. So perhaps there is a market out there for a baloney car treatment.

This little adventure was not without consequence. I was caught purchasing my baloney by a reader of my work. "How can you pretend to know about nutrition and eat that stuff?" he asked incredulously. I don't think he bought the explanation about my upcoming experiment and is probably out there spreading the rumor that I'm full of baloney. One other thing: it turns out that driving around with baloney taped to the car is not a good idea. There's no problem with the baloney, but the tape is hell to get off! (It takes a little WD-40 and elbow grease.)

Not every question I receive is as easily researched as the effect of baloney on car paint. One that came my way, about frozen goat ears, falls into the "more difficult" category. A woman who raises goats for meat was interested in an explanation about the effects of vinegar on the animals' frozen ears. It seems that if goats are kept outdoors in the winter, the tips of their ears can freeze and literally drop off. I doubt that goat ears have great commercial value; the concern is for the animals' welfare. If the frozen ears are dipped in vinegar, I was told, the tips stay attached and goat happiness is maintained. The only problem? The hairs that grow back on the previously frozen ear are white.

Experimenting with goat ears is a far more daunting task than decorating a car with baloney. After being told that goat ears, whether attached to the animal or not, were non grata in the house, I decided that my research had to take a different tack. The baloney, which I already had in my possession and now felt too guilty to eat, would perhaps provide a workable model for a goat ear. To expedite the proceedings, I froze half of a slice in liquid nitrogen. I'm well aware that the appropriate treatment for frozen tissue, as in frostbite, is slow warming. So I already suspected that the "vinegar effect" had less to do with vinegar and more to do with a room-temperature liquid being a good way to slowly warm tissue. Indeed, there was no detectable difference when the frozen baloney was warmed with vinegar, water, or Diet Coke. Of course, without a definitive experiment on real goats, I can only venture an opinion that when it comes to warming frozen ears, vinegar offers no advantage over other liquids. Since my baloney samples were incapable of growing hair, I have no idea why goat ear hair turns white after the vinegar treatment.

I have also had questions thrown my way that I am unwilling or unable to answer. For example: What does arsenic taste like? (You're not going to catch me sampling that stuff.) Is it safe to eat candle wax? (Probably, but why would you want to?) Is dehumidifier water safe to drink? (Probably—and it may wash down candle wax). Do you urinate more after drinking well water than tap water? (Probably not, but I've reached an age where increasing urinary frequency is not an attractive research proposition). Does using plastic cutlery impair male performance? (I suppose it depends on what you use it for.) Can you entice an octopus from its hiding place with a blue-colored rock? (If octopi want to hide, I'm all for leaving them be.)

Here's a final off-the-wall question: Can aspartame induce "man breasts"? A worried correspondent alerted me to a Web site that sports a picture of an older gentleman wearing a bra. The man in the photo explains that he has done a lot of research on this matter and has come to the conclusion that his breast enlargement was caused by aspartame. I suspect his "research" was limited to psuedoscientific drivel circulated by the vocal anti-aspartame movement, which claims that the artificial sweetener is responsible for virtually every ailment known to humankind and "completely destroys the brain." It seems, though, that the man's enlarged breasts were not without their merits. They allowed their owner to strike up frequent conversations with women about bra brands and sizes and, one assumes, aspartame and breast growth.

So what is my take on this phenomenon? Well, the effect on car paint is not the only question I've gotten about baloney.

A HOLE IN THE HEAD:
SCREWY SCIENCE

The camera closes in on the electric drill boring a hole through the skull. Blood gushes from the wound and splatters everywhere. The latest horror movie? No. This scene is from a 1970 "documentary" called *Heartbeat in the Brain*, a true cinematic classic. The message of the film is clear: If you want to achieve a higher degree of consciousness, you need a hole in your head!

In this case the hole belongs to Amanda Feilding, a British artist who, in the film, calmly shaves her head, makes an incision with a scalpel, and then attacks her skull with an electric drill. Joey Mellen, the cameraman, is a veteran of self-administered surgery and already sports a hole in his head. Why did these two decide to ventilate their brains in this fashion? Because they bought into a truly remarkable theory advanced by Bart Huges, a Dutch medical student who failed his finals and never managed to get his degree. Given his ideas about brain function, maybe we shouldn't be too surprised at Huges's lack of academic success. Plainly stated, the enemy of clear thinking, according to this Dutch "savant," is gravity. Much of human misfortune, he claims, can be traced to the time when the "ape first stood upright." We'll ignore for now the fact that while we may have evolved from a common ancestor, we did not—except perhaps for Huges—descend from apes. In any case, Huges contends that when humans became upright, there was a loss of blood from the brain because the heart had to propel blood against the force of gravity. Consequently, the brain had to take emergency measures to ensure that the parts essential for survival received sufficient blood. Capillary vessels leading to less important areas of the brain were constricted, reducing blood flow and, in consequence, the delivery of glucose and other nutrients. Thinking, it seems, is not one of the brain's more important functions. Luckily, however, there is a remedy for brain activity that has been impaired by gravity.

According to Huges, if a proper level of consciousness is to be reached, blood volume in the capillaries has to be increased. The most obvious way to achieve this increase is to stand on one's head and allow gravity to undo the mischief that it caused in the first place. By this reasoning, bats should be among the most intelligent of creatures. Jumping from a hot bath into cold water also does the trick. So do certain drugs, such as LSD. This substance, Huges says (without any evidence), constricts the veins in the neck, impeding the exit of blood from the brain, thereby making us smarter. But these are temporary measures. A permanent higher level of consciousness can be achieved by boring a hole in the skull, a technique known as "trepanation." The convoluted theory explains that if the brain in not constricted by the skull, the heart can more easily pump blood into it. Huges refers to babies' skulls, which are not yet closed and clearly show the "pulsation" of the blood in the brain with each heartbeat. What is not clear, however, is the relevance of this argument. Nor is it apparent why people who "unseal" their skulls for "psychic buoyancy" feel that their rationale for doing so is buttressed by the discovery of trepanned skulls around the world.

It is certainly true that archeologists have dug up skulls with holes in them. Some of these finds date back to the Stone Age. But these artifacts do not show that our ancestors fortuitously discovered the secret to enlightenment. They do show that some primitive peoples had surprising surgical techniques when it came to treating skull fractures caused by slingshots or clubs. Bone fragments were neatly excised to prevent brain damage. Careful examination of the bone structure around these wounds shows that in some cases, the bone healed and the patient survived. Perhaps in some cases such surgery was also performed as a treatment of mental illness by means of releasing "evil spirits."

Is there anything to the suggestion that a hole in the head is good for us? Brain function is actually related to blood flow, not blood volume. And drilling a hole into the skull does not increase blood flow. Many patients, of course, have had holes drilled in their skulls for various types of brain surgery, but none of them has reported any sort of enhanced mental clarity as a result. A typical example of this sort of procedure would be temporary removal of part of the skull after aneurysm surgery to accommodate swelling of the brain. In one fascinating Australian case, the surgeon misplaced the bone that was removed, and the hole was eventually covered with a titanium plate. The patient sued the hospital because she said she couldn't get over the feeling that part of her skull had been fed to a dog. Maybe if they had left the hole uncovered she would have reasoned with more mental clarity.

Just what kind of enhanced wisdom did Amanda Feilding achieve? It's hard to say. Not a great deal if you consider that in 2000, she felt the need to repeat the trepanation. But maybe she experienced a boost in intellect from the original hole, because she found a neurosurgeon in Mexico City to drill her skull the second time. You have to wonder, though, about the state of the surgeon's head. Ms. Feilding also ran for parliament twice (a hole in the head, of course, is not a disqualifier for political pursuits) on a platform that trepanation should be covered under Britain's National Health Plan. She garnered 49 votes the first time and 139 the next. Obviously there are a number of voters in Britain who should have their heads examined. And some on this side of the pond as well.

Peter Halvorson, a trepanned American, runs an advocacy group that promotes holes in the head. He does have some comforting thoughts, though. Apparently, in about 10 percent of the population, the intracranial seams that we all have as children do not heal and the natural openings provide for enhanced brain function. John Lennon, Halvorson says, was an example of this phenomenon (and not a particularly good one, I would think), as are physicians. It is physicians' positive pulse pressure that allows them to complete the extensive education required to become a doctor. I imagine that education prevents them from wanting to mangle their skulls. I don't know where Halvorson stands on chemists who write books. I'm pretty open-minded, but I don't want my brain falling out. So I'll leave my skull the way it is. Perhaps I'll try a few headstands instead, or jumping from a sauna into cold water. You be the judge—see if the following entries are more enlightened . . .

THE QUIRKS OF QUARTZ

When I was in elementary school in Hungary, students had to go through a seemingly bizarre ritual several times every winter. Stripped down to our shorts, we stood in front of a device that was always referred to as a "quartz" lamp. We really had no idea what this ritual was all about except that it had something to do with health. Obviously, it was an ultraviolet lamp designed to emit wavelengths that trigger the production of vitamin D in our skin—a process to compensate for lack of sunlight in winter. Why was the device called a "quartz" lamp? I had no idea then, but I know now. Regular glass absorbs a lot of ultraviolet light but quartz glass does not. So it makes sense that we had to cavort around in front of an ultraviolet source sealed in a quartz bulb.

What is quartz? Chemically speaking, it's a mineral composed of an organized three-dimensional array of silicon and oxygen atoms that correspond to the basic formula of silicon dioxide. It is a constituent of many rocks, such as granite, but can also be found in nature in the form of beautiful colorless crystals. These crystals have an amazing property: they are "piezoelectric." When subjected to an external electric field, they vibrate. In other words, quartz can translate an electrical signal into mechanical motion. This property makes for extremely accurate clocks as well as various electronic devices in which frequency control is critical. It also makes for a variety of nonsensical claims based on supposed therapeutic frequencies produced by a quartz crystal.

The usual inane claim is that some interaction between the vibrations of the crystal and the vibrations of a person's body will trigger various responses. We are "informed" that every cell in the body vibrates at its own specific frequency and disease occurs when these natural frequencies become unbalanced. Actually, if anything is unbalanced, it is this bizarre theory. In any case, the contention is that the situation can be corrected if crystals are placed, as one Web site claims, on "one or more of the major energy centers of the body." This technique purportedly eliminates harmful vibrations produced by environmental toxins or emotional stress. The Web site even promises that the appropriate placement of crystals on the body increases the size of the genitals. But more commonly it is a person's energy level that supposedly can be increased with crystals. Often a quartz crystal is placed in each of a patient's hands to promote the flow of energy and "balance the polarity between the right and left sides of the body." Undoubtedly, the patient will often feel better. But this result has everything to do with the power of suggestion and nothing to do with quartz vibrations. At any rate, quartz crystals only vibrate if they are exposed to an electrical field.

Unless, of course, they have been exposed to Uri Geller's "mind power." This extraordinary—at least in his own mind—psychic has made a career out of bending spoons and starting stopped watches. One would think that this fantastic ability to bend metal could be put to a better use than for tormenting cutlery. And now it seems it has. Geller has beamed his psychic energy into a piece of quartz, which is included in the Uri Geller MindPower Kit along with a book, a cassette, and a vibrantly colored orange dot that Uri says activates will power and psychic energies. The crystal is very small. But according to believers, its size doesn't matter. One erudite salesperson in a crystal shop tells potential customers not to worry about the small size of the crystal, pointing out that the crystals that power watches are even tinier. Funny, I always thought those watches were powered by batteries.

Geller's book tells us that his crystal is suitable for curing all kinds of ailments. "You can carry it in your pocket on days when you don't feel well, you can wear it all the time to boost your energy levels and ward off infections, or you can hold it in your hand and concentrate on it whenever you feel ill or lacking in energy." The only thing that is crystal clear to me about this stuff is that Geller has found another way to make lots of money.

Crystals intrigued people long before Uri Geller and his ilk made them into therapeutic commodities. The word *crystal* derives from the Greek for "ice." That's because the ancient Greeks believed crystals were some sort of permanently hardened ice. Indeed, this view persisted until the seventeenth century, when the famed chemist Robert Boyle laid this misconception to rest with a simple experiment. He studied the variety of crystal that had been christened "quartz" sometime in the fourteenth century by Bohemian miners, after the root of the Slavic word for "hard." Boyle simply measured the density of quartz by weighing a sample and immersing it in water; he found it to be almost three times as dense as ice. Obviously, quartz was a different species.

Quartz is clear unless small amounts of impurities such as iron or manganese get incorporated into the crystal structure. Then a variety of colors can ensue, the most intriguing of which is the purplish shade that characterizes amethyst. Amethyst was the most coveted form of quartz in ancient times based on the belief that it protected the wearer from the consequences of drinking. Indeed, the name "amethyst" comes from the Greek *amethystos,* for "unintoxicated." In ancient times amethyst was sometimes put into wine cups to protect the drinker from the effects of alcohol. Perhaps this strange custom was prompted by the wine-like color of amethyst. At least the amethyst didn't have to be consumed. Other therapeutic regimens, popular at various times in history, focused on adding powdered crystals to beverages and then drinking the concoction. The more rare the stone, the more potent the supposed effect. Obviously, this kind of "mineral water" was only for the rich. In the sixteenth century, Pope Clement VII is said to have ingested a fortune in precious stones.

We no longer grind up gems for consumption. I guess a few centuries' worth of negative evidence were enough to convince people that this practice just didn't work. But are crystals totally useless in protecting us from outside influences? Not if you ask the promoters of the BioElectric Shield, which is a pendant that contains quartz and other minerals in a specific pattern designed to shield the wearer from one of the horrors of modern life: electromagnetic fields generated by power lines. The pendant seems a bargain at $140. The ad promoting this miracle does caution that some people will feel a difference while others will not. Perhaps to improve its performance, buyers should get the quartz psychically energized by Uri Geller. I had better be careful what I say about Uri—I might end up with watches that don't work and a drawerful of bent cutlery.

The Sad Story of a Little Angel

Gently holding the hand of five-year-old Kathy Allison, John Haluska walked to the podium in the Pennsylvania State Senate. "Here is a little angel," he told his colleagues, "who, according to medical science, had to meet the angels soon. But after receiving the Hoxsey treatment in Dallas, she is going to school and is cancer-free. And they still call Harry Hoxsey a quack."

Haluska's dramatic little speech on that day in 1958 was aimed at eliciting senate support for a cancer treatment clinic his good friend Harry Hoxsey planned to open. Hoxsey needed the support. For over thirty years the American Medical Association (AMA) and the US Food and Drug Administration (FDA) had been on his trail, attempting to throw a monkey wrench into his well-oiled "cancer cure" machinery.

The trail was a long one and wound through West Virginia, Michigan, New Jersey, and Texas. Hoxsey had set up a string of clinics where he claimed to have cured thousands of cancer patients. Orthodox physicians, he maintained, mutilated their patients with surgery and burned them with radiation in a futile effort to destroy cancer. But he had found a way to restore health with gentle, herbal treatments. Of course, "the establishment" had turned against him because he was destroying their business. Doctors were making millions with their needless surgeries and radiation treatments and had no interest in providing cheap, effective therapies. Not only that, they were actually making people sick with their "cow pus" vaccinations and endorsements of preservatives in foods and "rat poison" (i.e., fluoride) in drinking water.

Hoxsey's story of his rise from rags to riches and his one-man struggle against powerful government forces played well to people who felt they were being harassed by excessive government interference and made miserable by corporate greed. His fiery rhetoric about evil monopolies, conniving Jews, and dastardly communists hit home with many Americans who were struggling to eke out an existence. For ostentatious orators like Hoxsey, a good scapegoat always comes in handy.

You Don't Have to Die was the captivating title of Harry Hoxsey's 1956 autobiography. In it, he described how his great-grandfather, a Kentucky farmer, had noted a cancerous growth on the leg of one of his stallions. The vet advised that the horse be put down, but farmer Hoxsey decided to put the animal out to pasture and let nature take its course. Remarkably, the stallion recovered. Hoxsey had noted that the horse always grazed in one particular area and concluded that the plants growing there must have been responsible for the miraculous cure. He then blended various parts of these plants to produce three specialized "cancer cures." The secret formulas were handed down and eventually put to use by Harry's father in treating cancer patients. While he claimed spectacular results, apparently the formula did not work for him; Harry's father developed cancer of the jaw and decided that conventional radiation was a better option. Harry denied his father's medical history and claimed to his dying day that the AMA had fabricated a false death certificate and his father had really died of an infection. In any case, the elder Hoxsey passed the cancer formula to Harry on his death-bed and warned him that "they will persecute you, slander you and try to drive you off the face of the earth." A savior was born.

The secret formula turned out to be a mixture of red clover, prickly ash, buckhorn, alfalfa, and potassium iodide. But according to Hoxsey, it was the specific blend and amounts used that were critical. "Bunk," said the AMA, which filed injunction after injunction. Hoxsey fought back. He was the victim of a conspiracy, he moaned. "Is it possible to sell a 'fake' cure to 10,000 people for 30 years, despite the vociferous opposition of organized medicine and still attract forty new patients a day?" he asked rhetorically. Actually, it is. And it's rather easy. Desperate patients will do desperate things. And it's hard to blame them, especially when traditional medicine is unable to provide the "guarantees" that Hoxsey did.

The healer's "successes" can be readily explained. Since he or his workers did the original diagnoses, it is a good bet that many of the patients never actually had cancer. Indeed, Hoxsey maintained that "any man who has to resort to a biopsy lacks experience or mistrusts his own ability." A former patient testified that he had been diagnosed with cancer and offered a treatment for $250 and a six-week recuperative stay at Hoxsey's hospital for $360—a lot of money at the time. He recovered. But not from cancer. It turned out that he had really been suffering from "barber's itch," or folliculitis, an irritation of the hair follicles. In another instance, an FDA undercover agent was diagnosed with advanced prostate cancer that had metastasized to the lungs and was told he had come to the clinic just in time for the cancer to be arrested. There was an arrest all right, but it wasn't of the cancer. Some of Hoxsey's patients certainly experienced a placebo effect and others proclaimed publicly that they had been cured, probably in an attempt to convince themselves. Nobody likes to admit to having been duped. As the saying goes, the plural of "anecdote" is not "evidence."

Hoxsey repeatedly challenged the AMA to investigate his "cure." "How can you condemn a treatment without studying it?" he asked. Of course, Hoxsey himself never initiated a study despite having become immensely wealthy and possessing the means to fund a proper controlled trial. The AMA accepted the challenge and twice asked Hoxsey to provide patient files. He did, but they were so devoid of proper medical histories and records of physical exams that they could not be evaluated. In 1999, the National Center for Complementary and Alternative Medicine (certainly not an anti-alternative organization) examined evidence submitted by a Hoxsey clinic in Mexico (yes, these clinics still exist there) and found that of 149 patients who had been treated, only 85 could be tracked down five years later and only 17 of these patients were still alive. This 26 percent survival rate is not exactly the claimed 80 percent rate and probably could have been achieved with an anticancer diet of frog legs, snails, and Mexican jumping beans.

Today, Hoxsey proponents wave scientific papers at skeptics. These papers contain data about the anticancer properties of some of the plants used in Hoxsey's formulations. These data are meaningless. There are thousands of plants that, in laboratory studies, show such properties and have no clinical relevance. Hoxsey himself is a testimonial to this fact. He developed prostate cancer and when he failed to cure himself, he quietly underwent conventional surgery. And what of Kathy Allison? Eight months after being "cured" by Hoxsey, she was dead of cancer. So contrary to John Haluska's remarks, evidence indicates that Harry Hoxsey purported to have medical knowledge that he did not actually possess. In other words, he was a quack. Which is precisely why the government finally managed to put him out of business by 1960. His legacy though lives on today in the antics of numerous bogus cancer-cure gurus who prey upon the desperate.

Coral Calcium on the Brain

How would you like to get a brand-new brain? According to "Dr." Robert Barefoot, it isn't that hard. All you have to do is bathe your old brain in calcium. Not any calcium, mind you, but "coral calcium." And not any old coral calcium either. There are hucksters out there trying to cash in on this miracle by scamming the public with poor-quality coral. But not our pal Barefoot. This people's champion knows the source of supplements that have the "perfect balance" of calcium, magnesium, and trace minerals. Only coral gathered from the ocean around the Japanese island of Okinawa has the right stuff for brain regeneration. And don't worry—you don't actually have to remove the organ from your skull to rejuvenate it; you can bathe it "in situ" by taking coral-calcium supplements orally. But a new brain is only part of the coral-calcium sensation. This wondrous product also offers protection against over 200 diseases, including cancer, arthritis, and Parkinson's disease. Needless to say, it is also ideal for weight loss and, of course, will turn gray hair back to its original color.

None of this is news to TV viewers who have seen Barefoot crop up on "infomercials" with annoying frequency. These programs are designed to look like documentaries and usually feature a "skeptical" interviewer who tries, unsuccessfully, to poke holes in the unconventional theories of an "expert" guest. In the case of coral calcium, the interviewer is a performer with a rather sketchy personal history that includes spending time in jail for check fraud. He has also had numerous legal encounters with the US Federal Trade Commission (FTC) for promoting a variety of nonsensical products on the air. Once he even posed as a doctor to improve his image with bank officials. The coral-calcium expert is, of course, our friend Robert Barefoot.

First things first. Barefoot is not any kind of a "Dr."—not an MD, not a Ph.D. His educational background consists of a degree in chemistry from the Northern Alberta Institute of Technology, a degree that should be revoked in light of the chemical arguments he makes on behalf of coral calcium. Arguments such as our body's DNA is virtually dormant if it is not "smothered in calcium" and if you are calcium deficient "you grow old twice as fast and your body can't repair itself." As is the case for most clever knaves involved in marketing schemes, Barefoot takes a speck of scientific truth and blows it out of proportion.

Yes, DNA replication does require calcium. And potassium. And iron. And magnesium. And a host of enzymes. And dozens of other biochemicals. Replication only happens properly in healthy cells and cellular health requires the correct balance of all of these substances. To suggest that calcium supplements alone cause DNA to replicate properly and the brain and other tissues to regenerate is absurd. Our body has a vast storage depot for calcium that can be tapped if needed. It's called the skeleton. Calcium deficiency can certainly lead to osteoporosis, but it will not wither your brain. Barefoot also sheds light on Parkinson's disease, which he says is caused by the accumulation of electron-deficient free radicals in the brain. These free radicals are "starved" for electrons, which can be provided by calcium. This is a truly bizarre notion given that in the body, calcium exists in the form of positive calcium ions, which most assuredly cannot give up any more electrons.

There's more: "Ninety-five percent of all disease can be cured in America in two years with coral calcium." According to Barefoot, doctors don't know this statistic because all their knowledge comes from the pharmaceutical industry, which has no interest in promoting "safe and natural" cures. That's why the medical establishment hasn't learned that heart disease is actually quite easy to treat. It is caused by acidosis, not cholesterol. Cholesterol is actually a lifesaver! It seals cracks in arteries caused by acidosis. Yes, Barefoot admits, cholesterol is found at the scene of the crime. But he asks rhetorically, "Are you concerned if firemen show up at a fire?" He goes on: "God gave us cholesterol to seal cracks in our arteries and without it we would be dead." Barefoot tells us that if those over the age of seventy double their cholesterol, they will live longer. He makes it sound as if we should be guzzling the stuff.

The self-proclaimed "renowned chemist" attempts to buttress his argument by quoting Dr. Kilmer McCully, a noted researcher who says that "acid eating a hole in the wall of an artery, and not cholesterol, is the cause of heart disease." McCully did propose a theory that an amino acid, homocysteine, is a risk factor for heart disease—but certainly not because it eats through arteries. In any case, to reduce homocysteine one needs to ingest B vitamins, not calcium.

Apparently, acidosis is also the cause of cancer. According to Barefoot, the acidity of saliva is a reflection of the acidity of the blood, and cancer patients have a salivary pH of 4.5, which is very acidic. It is also very incorrect. No studies have shown a relationship between the pH of saliva and cancer. But more important, the acid-base balance of the blood is not altered with diet. The blood is a buffer system maintained by the body in a very narrow pH range of around 7.4. The suggestion that this balance can be altered by taking coral calcium is ridiculous. If cancer were a question of adjusting body pH with diet, a dilute solution of sodium hydroxide, as found in clog remover or oven cleaner, should also get the job done.

Barefoot totally misquotes a study published in the *Journal of the American Medical Association* and claims it shows that calcium cures cancer. It shows no such thing. The study found that people who were at risk for colon cancer and who increased their dietary intake of calcium from dairy products—not supplements—reduced their risk. It is true that many North Americans do not get enough calcium and should consider supplements for various reasons. Calcium carbonate is a good supplement, but it doesn't matter if it comes from exorbitantly expensive coral-calcium products, or cheap house brands, or, for that matter, blackboard chalk. Want to hear more? Barefoot says he's seen multiple sclerosis patients leave their wheelchairs behind and thousands of patients cure themselves of cancer with coral calcium. He says that people on the brink of death come to him and recover after following his program. Or, he used to say such things. Now that the FTC has clipped his wings, he may be singing a different tune. Barefoot settled with the FTC and agreed to stop making claims that Coral Calcium Supreme or any other coral-calcium product can treat or cure cancer, multiple sclerosis, heart disease, high blood pressure, or other serious diseases. He is also prohibited from making unsubstantiated claims that coral calcium is more bioavailable (i.e., more easily absorbed) than other calcium supplements (Barefoot Coral Calcium Supreme was touted as providing the same amount of bioavailable calcium as two gallons of milk).

Presumably, Barefoot will also cease advertising that you can acquire a "new" brain by "bathing" your current one in coral calcium. Although, a new brain for Mr. Barefoot might not be a bad idea. Judging by all the foolishness he spouts, those pills are not getting the job done.

DEAD WATER AND LIVING QUACKERY

If you live in a major city, you probably have some worries about crime in the streets. But what about crime under the streets? Did you know that murder is being perpetrated there all the time, right under your nose? Water pipes are killing our water! I wasn't aware of this criminal activity until I was informed of it by the purveyors of the Original Danish Water Revitalizer, who are dedicated to bringing the crime wave to an end.

What is the cause of the problem? Modern water technology. It seems we force water to flow through straight and narrow pipes, allowing it "no freedom to follow its innate desire to move in spirals and swirls." We even expose the water to "deadly 90 degree turns." This, I'm told, is a particularly dastardly thing to do. As the unfortunate water molecules smash into the pipe, the bond angle between the two hydrogen-oxygen bonds is reduced, and if it reaches 101 degrees, the water "dies." And dead water, of course, has no energy to fight off bacteria and cannot service our bodies properly. Given that our bodies are composed of roughly 70 percent water, I guess we should not be surprised that there is so much illness out there.

Ah, but there is a solution. The Original Water Revitalizer will "give the water a double helix spiral which creates a vortex energy field and restructures the water at the molecular level." It is said to restore the water's energy in the same way nature does. Apparently nature accomplishes this task by having water flow through winding rivers. Needless to say, the Revitalizer also "removes from the water's memory the electromagnetic frequencies of any pollutants which are just as harmful as the pollutants themselves." All this is accomplished without any parts that wear out and without any filters that need replacing. Surely a bargain for $179!

And what do you get for your investment? Nothing more than a piece of curved copper pipe that can be attached to your faucet! All those dead water molecules, the victims of the straight and narrow pipes, will be resurrected as they cruise through the life-giving curve. The theological consequences alone are staggering! So are the testimonials offered by people who drink, wash, and shower with "living water." They feel more energetic, use less detergent, and no longer have to cope with the taste of chlorine. There must be some novel chemistry at work here, because my chemical education has not prepared me to explain how chlorine is removed by having the water flow through a curve. Nonetheless, some people are convinced. They say they are no longer affected by the chlorine in the water and are happy to have replaced their previous water filters, which they now realize were conspirators in the crimes against water. After all, those filters forced the water through their straight tubes!

So you think I'm making this up? Well, I'm not. I don't have that much imagination. And it really does take a great deal of imagination to assemble the kind of preposterous gobbledy-gook seen in the advertisements for the Original Danish Water Revitalizer. The scariest aspect of this little story is that obviously there are people buying into this scam. These people even think they are helping the environment; they are told the effect of the Revitalizer is so strong that when the water is transported back to nature it cleans the rivers, lakes, and oceans. Instead of attaching curved pipes to our faucets, we need to educate people about science so that they are equipped to straighten out the crooked purveyors of such nonsense.

Battling nonsense, though, is a tough task. Whenever you think you have seen the ultimate in absurdity, something else comes along and reaches even loftier heights. The Centre for Implosion Research in England is also out to publicize water's confrontation with straight motion. This group's remedy is not just using a curved pipe but also creating a "vortex": "Natural water forms whirls and eddies, and given the freedom, will form a vortex, as in going down the drain." Presumably, vortexes create great energy. Anyone with doubts about this claim is counseled to look at what a tornado can do. The center maintains that life lies in vortexes and refers to the spiral structure of DNA as proof. It seems it is easy to infuse vortex energy into water. And no, you don't have to buy a vortex-shaped pipe to do it. You do, however, have to purchase the Vortex Energiser, which is a spiral copper tube that contains "imploded" water. You simply place this contraption next to a water pipe or near any water that has to be brought back from the dead (putting it in your refrigerator will supposedly reenergize the water in fruits and vegetables that has been treated with "chemicals"). The "researchers" never explain what imploded water is; they only say that it is the opposite of "exploded" water, which I assume those of us unwilling to invest in the Vortex Energiser are drinking with great risk to our health. Maybe that's why I feel like exploding when I read about imploding water. The Energiser, filled with a lifetime supply of imploded water, goes for about $250.

Now, after all this balderdash about water pipes, let me mention a real water concern. Prior to about 1970, water pipes were made of cast iron or ductile iron without any internal protective liner against corrosion. After that date, pipes were protected, often by a thin layer of cement on the inside. The unprotected pipes have corroded over the years and have built up scale and rust deposits, often restricting water flow and reducing pressure. In some areas there is concern that in the case of fire, there would be inadequate water pressure to allow firefighters to do their job effectively. Furthermore, water that flows through corroded pipes comes out rusty, but that isn't the biggest problem. If bacteria get into the water, they can take up residence in the irregular crevices on the corroded surface and multiply, as apparently was the case in Walkerton, Ontario, a town in Canada that suffered an *E. coli* outbreak from contaminated drinking water. In communities with old pipes, residents are sometimes urged to boil their water as a way to prevent such outbreaks. Pipes can be rehabilitated by lining them on the inside with cement or epoxy. In the UK and the US this practice is being carried out systematically. In Canada, the city of Toronto cleans and lines some 170 kilometers of pipe every year. The province of Quebec only manages to clean and line a few kilometers every year. Relining water pipes is an excellent investment that pays dividends in the future. So it seems the people who promote the Original Danish Water Revitalizer and the Vortex Energiser are right about one thing: water pipes may be troublemakers. But the solution to the problem does not lie in their crooked advice.

GREAT MOVIE, JUNK SCIENCE

What is the link between emeralds, rubies, and Erin Brockovich? Simple. They all owe their fame to chromium. The minerals beryl and corundum are pretty boring unless they contain traces of chromium, in which case they're prized as green emeralds and red rubies. Erin Brockovich was a ho-hum legal assistant in California until she discovered that Pacific Gas & Electric had leached trace amounts of chromium into the water supply of the town of Hinkley. She was almost instantly elevated to the role of people's champion and parlayed the trace amounts of chromium into a $333 million settlement on behalf of a number of Hinkley citizens who, in her view, suffered a variety of ailments caused by chromium toxicity. This modern account of David versus Goliath was compelling enough to be made into a hit film starring Julia Roberts, which in turn propelled the real-life Erin Brockovich to stardom on the lecture circuit. Her paltry income as a legal assistant is a memory of the past, and Ms. Brockovich can now easily afford to adorn herself with chromium in the form of rubies and emeralds. But a troublesome question lurks in some scientists' minds. Were the riches and fame a reward for good science or good science fiction?

Michael Fumento, a lawyer and accomplished science writer, was one of the first to suggest that maybe the Empress had no clothes. And he didn't just mean it literally. Fumento dared to argue that this paragon of virtue, this heroine extraordinaire, this knight in shining armor who had slain the callous, uncaring giant corporation had successfully used emotion, not fact, as her weapon. This argument merits our scrutiny.

First, we have to learn a little about chromium. The metal itself is a lustrous, silvery substance that used to adorn our car bumpers before auto manufacturers decided to replace it with plastic (ensuring that even a minor collision will require replacement of the whole bumper). But what a difference three little electrons make! Just remove them from their orbits around the nucleus of a chromium atom and you've got a chromium-3 ion, which has completely different properties. This removal is exactly what happens when chromium combines with other elements to form compounds. And such compounds are always colored, which explains why the name "chromium" was derived from the Greek *chroma,* meaning "color." Chromium oxide, for example, is a beautiful green. If you want to see what it looks like, look for an American "greenback" in your wallet.

Chromium-3 is more than just the color of money. In tiny amounts it is actually required by the human body. It is needed to form glucose tolerance factor, a compound that enhances the action of insulin. Accordingly, chromium picolinate (a readily absorbable form) has been studied as a potential aid in the management of diabetes. Results are ambivalent. If subjects are put on a diet low in chromium, their insulin function becomes impaired and can be corrected with supplements. But there is no simple way to assess a person's chromium status and therefore no simple way to know who would benefit from supplements. Some studies have also suggested that blood cholesterol levels can be favorably affected with chromium supplements. The evidence is not compelling, but supplements in the range of 200 to 300 micrograms per day are safe enough. At doses around 600 micrograms, though, toxicity concerns arise. Chromium supplements are also heavily advertised for weight loss and muscle gain even though the preponderance of studies have shown no such effect. Indeed, the us Federal Trade Commission has stepped in on occasion and ordered manufacturers to cease making unsupported health claims for chromium-3 supplementation.

If you take away three more electrons from chromium-3, you have chromium-6, or hexavalent chromium. This is Brockovich's chromium and she is right—it is toxic. Hexavalent chromium is recognized as a carcinogen on the basis of an increased incidence of lung and nasal cancer among workers exposed to massive doses over long periods. Brockovich is also correct when she says that hexavalent chromium was used in the PG&E plant as a corrosion inhibitor and some of the substance was irresponsibly released into groundwater. She is still on a firm footing with her claim that groundwater levels of chromium-6 in the Hinkley area were elevated. But her lack of scientific training is blatantly exposed when she attempts to link every ailment in the area with the presence of chromium-6 in drinking water. Whether it's a miscarriage, a rash, bone deterioration, Crohn's disease, lupus, or any sort of cancer, Brockovich points the finger at chromium-6. In all probability she is wrong. Single toxins just do not behave in this fashion. They do not cause such a wide array of conditions. But when illness strikes, people are ready to pounce on convenient villains, particularly when there is potential for a large settlement.

Nobody contests the cause-and-effect link between inhaled chromium-6 and lung cancer. But ingesting trace amounts of chromium-6 is quite a different matter. The US Environmental Protection Agency (EPA) has examined the issue extensively and concluded that there is no evidence in the scientific literature to suggest that chromium-6 is carcinogenic by the oral route. This finding is actually not surprising to a chemist. It is well known that in the presence of hydrochloric acid, which is found in our stomachs, chromium-6 is converted into the innocuous chromium-3. The EPA's conclusion isn't surprising to an epidemiologist either. Studies have clearly shown that the cancer rate in Hinkley is the same as in the general California population.

So why did PG&E settle with the plaintiffs if there was no compelling scientific evidence to show that the myriad of symptoms suffered by the inhabitants of Hinkley were linked to chromium? Because the company was painfully aware that such cases are often decided on the basis of charismatic and emotional arguments instead of hard science. Erin Brockovich did a remarkable job painting a picture of a villainous, soulless, giant corporation that was heartlessly destroying lives for the sake of profits. Facts turn out to be a poor eraser for such an image. And who will the champion take on next? A worthy opponent like the tobacco industry? Unlikely. The heroine who proclaimed hexavalent chromium to be the smoking gun is a smoker.

BURSTING THE OXYGEN BUBBLE

Soothing music filled the air and dolphins appeared to leap out of the cascading waterfall that ran down the wall. I was guided to a comfortable chair and waited patiently as the attendant attached the cannula under my nose. My wife and daughter, who had been drafted to share in this adventure, were a little more apprehensive about the plastic tube poking into their nostrils. I suppose they were not quite as filled with the spirit of scientific discovery as I was. Finally, when the equipment was in place, we each pushed the little button on the control panel in front of us and the gas started to flow. The process we had come to experience was under way. We were being oxygenated!

I had wanted to try an "oxygen bar" ever since I first came across one in Toronto, Canada. I was captivated by the sign with its blunt statement that "You are what you breathe" and its promise to increase energy and stamina, rejuvenate cells and delay cell aging, increase mental alertness, deter fat, relieve migraines and hangovers, strengthen the immune system against viruses, and reduce environmental toxins in the body. It seemed so easy; these wonders would be performed by a gentle stream of oxygen directed into the nose. There were no pills to pop, no exercises to struggle with. Skeptical me, of course, wanted some firsthand—or first-nose—experience before discussing this scheme publicly. Now I had my chance and I sat there, inhaling, waiting for my batteries to be recharged. I even went the extra step: when the attendant asked if we wanted to try one of the bar's special "oxygenated fruit drinks," I said I was game. We breathed and drank, and then breathed some more and drank some more, and waited. Soon our time was up and our glasses were empty and we felt, well, the same as when we had sat down. I wasn't surprised. After all, I have studied a little physiology.

Our life really does depend on oxygen, which makes up 21 percent of the air by volume. And this has been so for at least 5 million years, contrary to what some oxygenation proponents suggest. One silly Web site says that in the 1800s, the atmosphere contained 38 percent oxygen and today, due to pollution and indiscriminate tree cutting, we have to cope with an inadequate oxygen supply. Bunk! Pollution does many nasty things, but it doesn't reduce atmospheric oxygen. Cutting down trees in a thoughtless fashion may have environmental consequences, but oxygen depletion is not one of them. The bulk of our oxygen supply is produced by algae in the oceans, not trees in the Amazon.

Why is oxygen so critical to life? Because the process of cellular respiration, the process that provides cells with the energy they need to carry out all the chemical reactions that constitute life, depends upon it. The key to such energy production is the removal of electrons from food components and their eventual transfer to oxygen, which can then combine with hydrogen ions to form water. This process happens in specialized areas of cells known as mitochondria. So how does the oxygen get from the air to the mitochondria? The journey begins when inhaled air inflates the 300 million or so tiny little sacks, the alveoli, that make up the lungs. Oxygen then diffuses into the blood that flows through the blood vessels lining the walls of the alveoli. Some of this oxygen dissolves in the watery portion of blood, the plasma. The amount that dissolves depends on how much oxygen there is in the lungs; the greater the oxygen pressure in these organs, the greater the amount that dissolves. But the solubility of oxygen in water is quite low, and the body's needs could never be fulfilled by the oxygen dissolved in blood.

Blood, however, also contains red blood cells, which harbor a complex molecule called "hemoglobin." Hemoglobin has the ability to bind with the dissolved oxygen. As it snares the oxygen molecules from solution, more diffuse from the alveoli to take their place. The process continues until each liter of blood contains the equivalent of about 200 milliliters of pure oxygen gas. About 197 milliliters, or 98.5 percent, is bound to hemoglobin, and 3 milliliters is dissolved in plasma. The heart then pumps this oxygenated blood to the body tissues, where the dissolved oxygen diffuses into cells and eventually into the mitochondria. As this happens, the dissolved oxygen is replenished by the release of oxygen from hemoglobin. Essentially, then, it is the storage capacity of hemoglobin that determines the amount of oxygen that can be delivered to cells. Now, the crux of the matter is that inhaling ordinary air, with 21 percent oxygen, leads to hemoglobin being over 98 percent saturated with the gas! Inhaling 100 percent oxygen would hardly make any difference. There would be a slight increase in dissolved oxygen, but this is of little consequence since the saturated hemoglobin is always ready to release oxygen into solution as needed. (Needless to say, those plastic strips athletes plaster on their noses with the hope of increasing oxygen intake by opening up the nostrils provide nothing but a placebo effect.)

This is not to say that inhaling extra oxygen is of no benefit. If someone has lung disease in which alveolar function is impaired, pumping more oxygen into the lungs can force more of the gas into the blood. Ditto for certain types of heart disease in which blood flow through the lungs is reduced. When the external oxygen pressure is low, such as at a high altitude, hemoglobin saturation is not achieved and breathing extra oxygen can help. In these situations, however, the gas has to be delivered continuously.

Now you can see why I was not surprised that inhaling a little extra oxygen for a few minutes had no effect. What about the oxygenated beverage? Well, it tasted pretty good, but there is no mechanism by which dissolved oxygen can enter the bloodstream from the digestive tract. Even if it could, it would be irrelevant. The solubility of oxygen in water is about 3 milliliters per 100 milliliters. A 200-milliliter beverage would then contain about 6 milliliters of dissolved oxygen. Since the body's tissues use about 250 milliliters of oxygen every minute, the oxygenated beverage would meet our needs for about a second and a half— if it could get into the blood. Of course, the normal path of entry for oxygen into the bloodstream is through the lungs, not through the stomach. Still, I'll admit that the experience of sitting in a comfortable chair, listening to calming music, watching a cleverly illuminated indoor waterfall while sipping a tasty fruit beverage does have its relaxing moments. But I could have done without the plastic tube sticking up my nose.

INDEX

About the Author

Joe Schwarcz is a professor of Chemistry and the Director of the Office for Science and Society at McGill University in Montreal, Canada. He hosts a popular weekly phone-in radio show, makes numerous television appearances, frequently gives entertaining and educational public lectures, and writes a column for the Montreal *Gazette*. He has received many honors, including the American Chemical Society's prestigious Grady-Stack Award for Interpreting Chemistry for the Public.